Beyond World-Class Productivity

Shigeyasu Sakamoto

Beyond World-Class Productivity

Industrial Engineering Practice and Theory

 Springer

Dr. Shigeyasu Sakamoto
Productivity Partner Inc.
140-1220
Maruyama 2
Nara 631-0056
Japan
DZV05747@nifty.com

ISBN 978-1-4471-6067-0 ISBN 978-1-84996-269-8 (eBook)
DOI 10.1007/978-1-84996-269-8
Springer London Dordrecht Heidelberg New York

British Library Cataloguing in Publication Data
A catalogue record for this book is available from the British Library

Cover design: eStudioCalamar, Figueres/Berlin

Printed on acid-free paper

Springer is part of Springer Science+Business Media (www.springer.com)

*This book is dedicated
to my parents with gratitude,
to my wife with love
and
to our children with hope.*

Preface

I've been concerned about the practice of industrial engineering for 4 decades. It's not easy to find useful books introducing the effectiveness of industrial engineering (IE) practice as it relates to the fundamental background of the field, its techniques, and in-depth theory. At one time, there was an abundance of useful books on motion and time study; however, the shelves display limited titles today.

Many books provide an overview of productivity and profitability as enlightenment for management, but these guides are not suitable for the practice itself in companies by professional engineers and their support staff. You will see plenty of titles defining useful technologies for inventory and lead-time improvement or participatory management practice, but it's not easy to find books concentrating on labor productivity that introduce basic tools of industrial engineering that can be applied in various industries.

Allow me to draw your attention to a discussion by consultants and professors many years ago in the *Journal of Industrial Engineering*. One of the key points was the introduction of classic IE, or modern IE. The age of computer technology came to IE in the form of new applications in work measurement and line balancing; mechanization, or automation, was set to transform manufacturing. Implementation of small group activity (SGA) and lean production entered many companies. As results were glorified regarding productivity and cost reduction, not only were terms associated with motion and time study virtually eliminated, industrial engineering itself became lost in translation.

These conditions were especially evident in Japan. Personally, I thought the classification "classic IE" or "modern IE" was not a suitable term. I preferred "basic IE". In the journal, Dr. Harold B. Maynard stated: "I do not for a moment believe that traditional industrial engineering is on the way out. Man did not discard the hammer when the saw was invented. He needed the hammer for pounding and the saw for cutting. In the same ways, IE needs different tools for solving different problems. He needs the old techniques as well as the new ones."

One reason that industrial engineering is in the shadows is that it is not known for contributions to management requirements. It may not get the trust by management due to its humble contribution, considering the many and hard requirements of true management.

There is an expression in Japan: "Gold coin for cat". A cat does not realize the value or usefulness of a gold coin. It has no meaning for a cat. With no value placed on it by the cat, the coin has nothing to do until the right person comes, attaches value and knows how to use it. In Japan, this is mind innovation; the right mind makes reasonable answers and attaches reasonable meanings.

At times, industrial engineering performs the activity of "nonreal gain", or small improvements with a small-cycle time reduction from time to time, place to place. The effect of this "improvement" is calculated by reduced cycle time in an annual occurrence. Such a calculated effect is a kind of ghost...invisible. Does this make sense?

Real gain should be pursuance. For example, reducing the allocated number of workers to reduce paid-hours immediately but accrue the same or more powerful results. This is an example of "real gain". Management, particularly in human resources departments, is interested in these types of gains. Industrial engineering should be a department that fosters these connections. Industrial engineering tools are effective enough to support management with these goals in mind.

Industrial engineering staffs should be cherished by management, given reasonable demands of improvement and receive them warmly. The result is that industrial engineers gain confidence and are motivated to develop higher standards of meeting staff services.

There are a lot of fashionable topics in productivity improvement, and there always will be. However, management and industrial engineers together must look ahead always. Basic industrial engineering technologies are not hackneyed. Effective results come when industrial engineers know how to use the technologies and demonstrate their abilities. This includes going back to the basics.

Experts never choose the tools themselves; as demonstrated in the following chapters; they need only apply them correctly.

Part I, Strategy for Improving Profitability and Productivity, introduces an overview and summary concerning significant points that management should care about in profitability and productivity. They should be eager to follow effective approaches not only in the interest of lean production but also participative management. There is a misunderstanding that if strong-market or high-profit companies are productive, there aren't many changes to make in the ways they do business. Strategy for manufacturing is not common but recommended in the interest of successful competition. Guess again. Companies must understand that there is a gold mine of productivity tools found only in a slightly different approach. The next three sections are filled with examples.

Part II, Theory of Productivity, presents a reasonable and precise theory about productivity. What is the true definition of productivity? Why is it important? International competition in today's business sphere is giving meaningful answers that readers can learn from.

Part III, Outline A of the Engineering Approach to Productivity, classifies productivity in three distinct dimensions that are particularly important to companies that desire large-scale improvements. What is the engineering approach that is effective in getting unique results? What is the difference between kaizen and the engineering approach in this book? The approach leads to nonempty gain. Methods engineering and searching for an innovative change of methods is key.

Many people are interested in productivity but misunderstand the relationship between corporate results and the approach to profitability and productivity. For example, the majority of kaizen or incremental improvement activities in manufacturing yield empty gains that do not stand out in business results. What is needed is a design approach focused on finding creative ideas that set and achieve theoretical design targets and directly impact earnings. In these chapters, I will present not only the core concepts of productivity improvement, but also a concrete approach for lasting success based on experience and results.

A concept of methods engineering that is not common in the world is introduced. Common sense and concrete contributions to corporate processes are described in this section. Additionally, work measurement practices are introduced with accurate, classical applications, but effective engineering for large contributions to improving productivity and profitability. A unique and practical approach based on engineering for challenging white collar productivity improvement is also introduced.

This book is the first time that some of this information will come to light. Improving "white collar areas" of productivity are also introduced.

Part IV, Monitoring Productivity, introduces fundamental ways to measure using theory. Means of measurement on the shop floor and in office areas are based on long-time consultancy-supported experiences.

Part V, Keys to Success for Improvement Management, provides cases. A company is required to restructure its organization, and the project team must concentrate on specific key indicators. Ordinary, or regular, attitudes and behaviors are not good enough. Mind innovation is required to successfully improve productivity and profitability. The single objective is to find the answer to what it is, not how to do it. I believe that any effective management tools are tools for stimulating mind innovation for the entire organization, and the right activities have to follow.

I am a management consultant with 40 years of experience in Europe, Asia, and Japan. This means that all the contents of this book are practiced with industrial engineering theory as the foundation.

Acknowledgements

I wish to acknowledge many people who gave me the opportunity to consult for their businesses and who supported me throughout the process of developing the ideas for a consultancy business. There are too many names for the size of this book to write down. Methods design concept (MDC), for example, was not developed and brushed up on without the help of my clients. Those companies are Ovaco (SKF) Steels, Volvo Car, Mölnlycke, Duni, Pripps, Electrolux, (Sweden), Spicers (UK), Whirlpool (Germany & Italy), Bekaert (Belgium), Korean Heavy Industry/KHIC (South Korea), Nippon Sheet Glass, Meiki, Nikon, Glico Foods, Toyota Body, Sony, Topy Industry, Nippon Fishery, Nippon Aluminums, Hoshizaki Electric, Deli Fresh Foods, and others.

Mr. Shirou Ishikawa, who worked at Mitsubishi Electric Co. Ltd., was a founder of my position on industrial engineering today. I could not have started my career as an industrial engineer without him; he gave me so much knowledge and experience in industrial engineering at Mitsubishi Electric, and agreed to change my job to a management consultant.

Mr. Takayoshi Nakajima, Executive Vice President at Japan Management Association (JMA), got me started as a management consultant at the Association. Mr. Takeji Kadota, Chief Vice President of the Association, trained and inspired me not only to develop and practice new management tools in industrial engineering, but also as a professional consultant to work to meet a high level of expectations by clients.

My international career regarding methods-time measurement (MTM) began with meeting Dr. Fred Evans from the UK MTM Association in Sydney, Australia, where I shared a banquet table with him at the International Conference of the World Academy of Productivity Science. He introduced me to Mr. Klaus Helmrich, General Secretary of the Swedish Rationalization Federation, in Sweden when the JMA required training for the new MTM-2 system; thereafter, I was trained and qualified through examination as the first international MTM instructor in Japan by the International MTM Directorate. Mr. Helmrich worked with me when I had the opportunity to consult on MDC for companies in

Sweden, UK, Germany, and Italy. He also introduced MDC into many other European countries. MDC was recommended as a chapter topic for the Maynard Industrial Engineering Handbook (fourth edition) by Mr. William M. Aiken, President of H.B. Maynard and Company and consequently my discussion of MDC forms Chapter 3.

My first global consultancy experience was provided by SKF Steel, and then the Volvo Car Corp. for implementing MDC in their productivity improvement. Mr. Dan Blomberg remembered my support and results from MDC for their change-over-efficiency projects at SKF Steel and then invited me to Volvo. Mr. Berndt Nyberg was a consultant in Finland; he gave me a few opportunities in Finland to introduce MDC. Because of those experiences I was invited to be a vice president by the President of Maynard MEC, Sweden, Mr. Lennart Gustavsson. I also express gratitude to my colleague consultants Mr. Takenori Akimoto, General Secretary of the Japan Institute of Plant Maintenance, Mr. Shouichi Saitou, Chief Vice President of the Japan Management Association Consulting, and Mr. Hideyuki Ueno, Chief Vice President of the Japan Management Association Consulting, with whom I trained in the Mento-Factor system in Den Haag, Netherlands. Gratitude also goes to Dr. Akihisa Fujita, Professor Emeritus of Kansai University in Japan.

I also want to acknowledge that I have benefited from input from the following individuals: Mrs. Joan and Mr. Allan Stuckey, Australia, Dr. Krish Pennathur, India, WAPS., Dr. James E. Lee, College of Business at Ohio University, Dr. James L. Riggs, Oregon State University, Mr. Robert E. Nolan, Robert E. Nolan Company Inc., USA., and Mr. Karl-David Sundberg, President of SKF Steel, Inexa, Sweden.

I also would like to express my special appreciation to Ms. Candi Cross, Managing Editor at the Institute of Industrial Engineers. Publishing an English book is a very unusual matter for the Japanese, but the endeavor has succeeded. Publication of my book has been completed successfully thanks in part to her support for my draft.

Then, I would like to thank Mrs. Nathalie Jacobs, Mr. Anthony Doyle, Mrs. Claire Protherough, and the entire production team at Springer UK, Mrs. Clare Hamilton, and Mrs. Sorina Moosdorf.

Finally, this book is a record of my 40 years of experience as a management consultant, especially in productivity. I want to thank my entire family for many different opportunities to share and support my consulting business. The business of consulting requires leaving home often, which causes of a lot of inconveniences in daily matters. This was made possible especially with my wife Kiyoko's support. Thank you to my daughters Koh and Masa for rewriting my English draft using their experience as graduates of Mills College, USA. This gave me the important opportunity to discuss my ideas of publishing a book in English regarding MDC and work measurement with my son Yuji, who earned an MBA in the graduate school of Nottingham, UK. He works as a Vice President of an international consulting firm in the field of management strategy.

With respect to a better way,

Nara, Japan   *Shigeyasu Sakamoto*
The Chrysanthemum Festival, 2009

About the Author

Dr. Shigeyasu Sakamoto is a management consultant in productivity improvement and president of Productivity Partner Incorporation. Before his current appointment, Sakamoto was Vice President of Maynard MEC AB (Sweden) and Vice President of JMAC (Japan).

Sakamoto is a Fellow at the World Academy of Productivity Science. He received his doctorate degree of policy science from the Graduate School of Doshisha University in Japan and is certified as a P.E. by the Japanese government. He is also certified as an industrial engineer from the European Institute of Industrial Engineering, International MTM instructor from International MTM Directorate (IMD) (1985), and a MOST instructor from Maynard Management Institute. He worked for the IMD as the technical coordinator responsible for developing a new

Photo by Mrs. Kiyoko Sakamoto

system of MTM. Sakamoto is a senior member of the Institute of Industrial Engineers. He's a Work-Factor and Mento-Factor instructor for WOFAC Corporation.

Sakamoto has published many books and papers in English and Japanese regarding productivity, industrial engineering, and work measurement. Recently, he explored the subject of company dignity through experiences as a management consultant of productivity for more than 20 years in Europe. He has questioned the quality of working life in Europe and Japan, distinguishing the habits of companies seeking big market share vs. those who strive for a culture of ethics and dignity. This study motivated his doctoral degree in research. The dissertation was published as *A Study of Company Dignity (SHAKAKU, Companality)*.

Contents

Abbreviations

AF	Auxiliary function
APT	Approval processing time
BF	Basic function
BS	Brainstorming
C	Combine
DLB	Dynamic line balancing
DTS	Direct time study
E	Eliminate
FM	Foreman
FS	Feasibility study
GDP	Gross domestic product
GNP	Gross national product
ICT	Ideal cycle time
IMD	Institute for Management Development
IMD	International MTM Directorate
IP	Input
ISEW	Index of sustainable economic welfare
IT	Information technology
JIT	Just in time
JPY	Japanese yen
LMS	Lowry, Maynard, and Stegemerten
M	Method
MBM	Measurement/monitoring base management
MDC	Methods design concept
MDH	Machining data handbook
MDW	Measured day work
MOP	Managing office productivity
MTM	Methods-time measurement
OHP	Overhead projector
OP	Output

OPM	Operational productivity measure
P	Performance
PC	Personal computer
PTS	Predetermined time standard
QCC	Quality control circle
QWL	Quality of working life
R	Rearrange
ROA	Return on assets
ROI	Return on investment
SAM	Sequential activity and methods analysis
SAMUS	Society for Advancement of Management
SEI	Stockholm Environment Institute
SGA	Small group activity
SLB	Static line balancing
SMED	Single minute exchange die
SOP	Standard operation procedure
ST	Standard time
STD	Standard time data
TCT	Target cycle time
TMU	Time measurement unit
TPM	Total productivity measure
TPS	Toyota production system
U	Utilization
UAS	Universal analyzing system
ULC	Unit labor cost
USD	United States dollar
WC	Work count
WCM	World-class manufacturing
WF	Work factor
WIP	Work in progress
WS	Work sampling
WU	Work unit

Part I
Strategy for Improving Profitability and Productivity

Chapter 1
Changing Strategy for Productivity and Profitability Activity

Chapter 1 provides an overview and summary of significant points that management should care about in profitability and productivity. The concept of "real gain" vs. "nonreal gain", for example, is introduced and encouraged as a condition for total improvement. Management should be eager to follow effective approaches not only in the interest of lean production but also participative management. There is a misunderstanding that if strong-market or high-profit companies are productive, there aren't many changes to make in the ways they do business. How will the next generation of engineers deal with declining profit margins? In the dawn of making sustainability a part of corporate culture, how can executives adjust to creating a reputation of corporate dignity rather than just economic growth? ULC is a useful measure to evaluate labor costs and productivity. This means international competitiveness on cost or price does not depend only on the level of wages but also on productivity.

1.1 Is Japanese Productivity Really High in World Competition?

What records are there in the world that tell the story concerning productivity? An international comparison is not easily achieved. For example, so-called Lean production, developed in Japanese car manufacturing, is copied and implemented all over the world as a specific remedy for productivity improvement. However, it is difficult to find reports or case studies that describe lean production's contribution to specific companies regarding productivity. Two examples illustrate this problem.

The first example is a field study of a production line of a Japanese company that produces car parts; the company belongs to the Toyota group (Spear and Bowen 1999). The name itself signifies that the company is believed to have implemented lean production well. The report shows a 208% productivity improvement over the

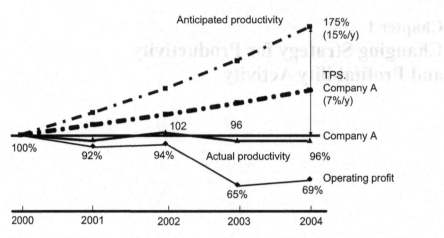

Figure 1.1 Declining productivity for 4 years: company A (From Industry Week 2007 Data: Harbour Group)

course of 11 years; it also shows approximately 7% per year in a production line. There is no way to know the percent of improvement for the company or a specific plant. This is just a guess, but the level of plant productivity could be lower than 7% annually. The level of productivity improvement is not reflective of a company that sincerely aims to raise productivity. Their annual rate of productivity improvement is not high enough in percentage to compete with other Japanese manufacturers.

Having less than 5% of improvement demonstrates a poor level of change; 10% is an acceptable lower level of change, and 15% is expected to be a leading company concerning productivity improvement. One of the best productivity improvements from a company was roughly 25% for several years. Companies with organized special teams for improvement projects have shown more than 200 and 500% within 2 to 5 years, and this is not unusual as you will see from case studies in this book. Note that the measuring method of your company does not matter (total production value in money or the number of production volume divided by the total consumed man-hours and so on). What ultimately matters is the end result.

How much productivity improvement has been executed so far that challenged the world-class manufacturing (WCM) level of productivity?

The second example is Figure 1.1, which shows the record of a company, A. The record of productivity, which was examined at the starting point of a project, then showed a 4% decline over 4 years. They had been doing active improvements using kaizen for a long time. Each kaizen appeared to be effective at the manufacturing department's level of productivity; however, those results did not contribute well at the division or company level. The company had anticipated a 175% productivity level if they could increase productivity by 15% per year. The company's new president hated such a reduction of production time, for example, and ordered a reduction in manning the improvements. Otherwise, nothing would have happened to improve the company as a whole.

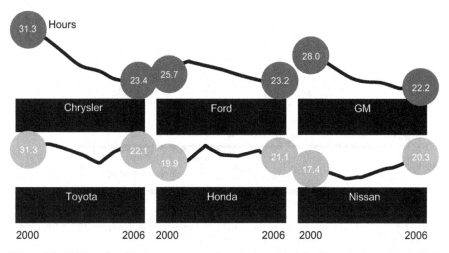

Figure 1.2 US-based and Japanese car manufacturers' productivity (From Industry Week 2007 Data: Harbour Group)

Figure 1.2 was introduced in *Industry Week*. The chart shows an interesting point among US-based and Japanese car manufacturers' productivity with average assembly time per vehicle from 2000 to 2006. The "Big Three" in the US declined 10–25% while the "Japanese Three" increased by up to 18% on assembly time per vehicle. Note that the Japanese Three are not as advanced today as you would think from this chart; compared to the US-based manufacturers, the Japanese manufacturers should be honor students in the lean production system, but they are behind more recently. The lean production system does not directly contribute to labor productivity improvement. It is effective in reducing cost. For the past 4 years, the Japanese manufacturers have not done a good job of improving their productivity. The analysis is not easy to do, but they used to be higher than the Big Three. However, today, this may be a myth.

Jim Frederick, a Tokyo-based writer for *Time* (Dec. 2002) magazine had harsh words for Japan's level of productivity in the article "Going Nowhere Fast": "Japan's labor force is one of the most unproductive of the industrialized world." He went on to cite statistics by the Japan Productivity Center for Socio-Economic Development, comparing workers in other countries as being up to 40% more productive than their Japanese counterparts.

1.2 Constantly Declining Profit Margins

The operating profit of Japanese companies' profits rose as sales grew from 1960 to 1980 but from the 1980s onward, profit margins declined even as sales expanded, and in more recent years, margins have even remained flat at the level to which they had fallen.

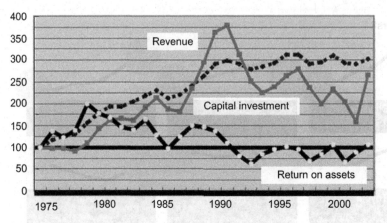

Figure 1.3 Capital investment, revenue, and ROA in Japanese Manufacturers: 1975=100 (from White paper, Japanese Economy, 1997) (Sakamoto 2001)

Behind this trend lies the intensification of international competition accompanying the globalization of corporate activity; that is, the advantages of the relatively lower labor costs in developing countries have led to the primacy of cost competitiveness. In considering cost competitiveness, we need to consider what proportion of cost is accounted for by materials and that, for most companies that import the majority of their materials (apart from certain ones), the country of their location is not going to offer advantages over a company located in other countries, so long as exchange rates do not skewer the cost of the imported materials. What remains as the cost competitiveness-defining item in the cost structure of goods sold or manufactured is mostly direct or indirect personnel costs. Similarly, personnel costs account for the majority of the cost of materials and purchased goods as well.

In other words, to enhance cost and price competitiveness, management must now work to effectively control the factor in the cost of manufacturing that it can control, which generally means controlling human resources and related costs.

Sales, profits, market share, margins and other similar numbers are generally of interest to managers as indicators of their company's performance. Profitability is yet another indicator of a business's health. Consider ROI in manufacturing industries. This is also known as the Du Pont Formula, named for the company that recognized its usefulness in management and introduced it to other companies around the world.

Of the many management indicators, which ones are really important for companies as measures of their performance? Managers are often strongly interested in sales, which make a company bigger, and naturally, the effectiveness of sales growth cannot be denied. The same goes for market share and profit margins. But managers need to consider whether it might be better to focus on indicators that are controlled by factors external to the company. Management's interests should be directed more toward indicators that reflect the quality of efforts being made within the company on so-called managerial items, factors that management has some control over. Profitability is just that powerful and effective managerial item.

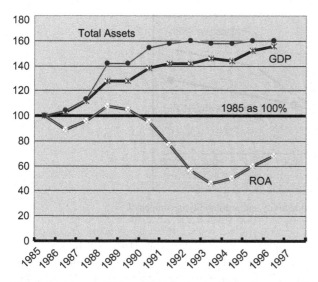

Figure 1.4 Assets, GDP and ROA in Japanese Manufacturers: 1985=100 (Sakamoto 2001)

Figure 1.3 (Sakamoto 2001) shows trends in revenue, investment in tangible assets (property, plant, and equipment), and ROA for typical Japanese manufacturing industries with fiscal year 1975 as the benchmark year. Until around 1990, an extremely strong correlation could be observed between growth in revenue and capital investment, although this weakened from then onward. In contrast, ROA has been declining since the beginning of the 1980s; in other words, even as sales have expanded, profitability has taken an inverse correlation and has continued to decline.

Figure 1.4 (Sakamoto 2001) depicts that for Japan, gross domestic product (GDP) followed increasing total assets (capital investment) and at the same time, there is a high correlation between capital investment and GDP. That means that capital investment contributed well to increasing GDP. GDP is equal to sales turnover or the production value of private companies. On the other hand, ROA declined even as GDP increased. Capital investment had put a lot of pressure on ROI causing it to decline. What is the purpose of company management, sales value or ROI and profitability? The key reason for declining ROI is capital investment for productivity improvement. Production volume increased in each manufacturer but profitability declined.

1.3 Potential for Major Profitability Increases

Figure 1.5 shows that company C achieved a 732% increase in profitability over 6 years in its production department. This rise by 7× the existing profitability was achieved on the back of dramatic improvements in productivity, with an improvement of approximately 510% over 6 years in manufacturing divisions that

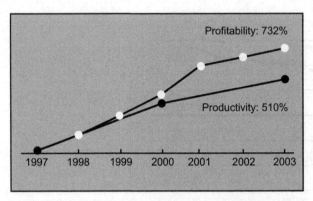

Figure 1.5 Improvement of productivity and profitability in company C

were subject to profitability improvement measures and 216% for the company overall, including business units not subject to the profitability enhancement measures. Although there was a huge change in sales figures over the 6 years, the outcome was not great; the improvements in productivity that occurred without capital expenditure also had a strong effect. This was achieved by management making rigorous decisions on capital expenditure aimed at boosting productivity and reducing costs, in principle, refraining from introducing new facilities that would entail investment and its relevant heavy costs. This illustrates how important it is to push forward with productivity improvements and cost reductions without spending money, or making investments.

Before it undertook productivity improvement measures, company C had 164 employees in manufacturing; as a result of these measures, it had 70 employees while maintaining the same production capacity, a reduction of 94 (Figure 1.6). The result demonstrates examples of workforce reductions at a range of companies that have implemented similar measures.

At a plant of a company, the workers were performing well and their working pace was satisfactory. The plant also had a synergistic approach to making efforts on a daily basis to reduce the product cost. An active innovation was being implemented, and I gave them credit in this respect. However, taking a closer look at the actual work, a worker used one hand only to hold the in-process product while the other hand repeated the same simple task. There appeared to be no special equipment for it and the workers were only working inefficiently.

Let us question: why aren't the supervisors and staff engineers interested in improving the work effectiveness per hour without taking into account the workers' individual efforts? For instance, complete use of both hands for tasks? Processing one product with both hands and processing two products in one cycle are possible.

Development of such effective work methods must be completed in advance of the instruction to workers. I don't mean to undermine the efforts of workers' independent improvement, but it is important to develop the right method from the productivity viewpoint in advance for the workers to practice company wide.

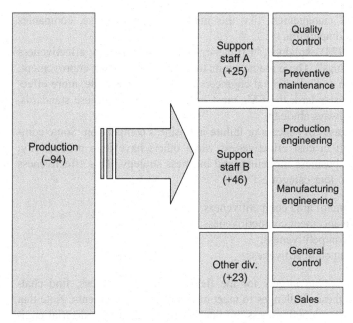

Figure 1.6 Utilization result of redundant areas: company C

Then there must be ideas from each worker for better improvement and a continuation of implementing the ideas. Set up the correct methods and practice faithfully; right things do right. Then you will see a good balance between the effects of industrial engineering and improvement based on the individual worker's talent and drive.

Even though the individual's effort toward improvement is important, I must emphasize the importance of strengthening the ability of the experienced supervisors' management.

Consider the example of a company in Japan whose production line began to manufacture digital cameras. They started this a month ago, but the line cycle time has not reached the plant's target. Then what happened at the shop? A staff engineer and foreman (FM) discussed ideas centered on reducing cycle time on the line side, and then they found that reach and movement on an electric driver was possible to reduce the time needed to perform the assembly. They were proud that the line cycle time met the plant target time. What comes to mind about this occurrence?

While I do not deny the improvement effectiveness, the action itself brought a very primitive level of improvement. It is possible to set up an effective method by having industrial engineers use standard measures that optimize principles of motion economy. The improvement that was made at the shop resulted from imperfect process planning. Those two employees were just doing trial and error due to the ignorance of pure basic industrial engineering techniques. The best way to meet the target cycle time would have been to prepare workers for the assignment.

Poor recognition of management like this must change; otherwise, companies cannot meet tough competition.

Peter Drucker (1973) explained, "Efficiency is doing things right; effectiveness is doing the right things". There are no limits to the possibilities for improvement. We never reach "the end". Industrial engineers should set reasonable, more effective methods as the standard; the FMs must instruct workers on these standards, which they should always abide by.

There are many cases of success or failure in business competition. Some companies have strength in cost effectiveness while others have it in productivity, product innovation, business modeling, and business strategy. This effectiveness can be classified into four categories:

- business model and/or area competitiveness;
- technological advancement competitiveness;
- low-cost ability competitiveness;
- active price setting competitiveness.

Industrial engineers set targets for the fields of competitiveness, find challenges, and resolve these challenges to meet management requirements. Note that the first two areas of competitiveness are not usually handled by industrial engineers, but rather corporate strategists. Low-cost tasks and active price setting are, however, within the skill set of industrial engineers. Still, note that company performance does not just depend on industrial engineering even though their abilities and experiences prove fruitful in nearly every industry.

Industrial engineering work is not about carrying out an attack in matters of competitiveness but rather defending high productivity and cost efficiency. Large-scale capital investments in full automation, for example, would be a decision by top management rather than industrial engineers. Industrial engineers attempt to find solutions for meeting management requirements without a large amount of expenditure. Active price setting competitiveness without low-cost conditions based on high productivity cannot lead to superior results for the company.

1.4 Enhancing Corporate Dignity Rather than Economic Growth

Herbert Stein introduced a new term in 1971: productivity (Stein 1971). Beforehand, the term efficiency was used in the place of productivity. The meaning of efficiency was to decrease the utilized labor rather than increasing the production volume. Eventually, productivity became not only dependent on efficiency but effectiveness as well. In addition, productivity includes a broader notion than conventional efficiency. People became interested in productivity combined with the quality of working life (QWL), which is comprised of: reduction of work hours, elimination of dirt and environmental hazards, and better living standards. From a company's point of view, productivity is to save labor, reduce cost, increase production, and beat competition. Given the two perspectives, a few questions are relevant. What conditions of productivity does management insist on? Can we envision an extension of the world market with a high level of productivity? What happens if a company climbs to the top position in the world market?

Management knows the appropriate answers. A better way will build up society. However, management cannot exclude the viewpoint of a company or not consider the economic advantage. Good management is interested in compatibility domains such as economic and social items. Let's introduce the importance of changing company policy to compatibility of social and economic domains simultaneously over just the economic points. I call this "company dignity", or "SHAKAKU, companality (a word I coined)". The bottom line is that productivity with higher profitability is the foundation for developing company dignity. Plenty of companies have succeeded on economic orientation only (through global expansion, for example). In contrast, many companies do not succeed on social orientation such as soft-skills maturity and integrity. Let's consider a few examples.

1.4.1 Changing from Growth to Maturity

Companies have been interested in increasing their sales and enlarging their market share, both of which are possible through economic indexes. Companies compete on their growth value in comparison with results of the previous year's position among competitors. Generally, business magazines inflate their growth through annual reports rather than their level of maturity. One of my concerns is why companies need growth and what size of growth results would be the final goal. It appears that management is not concerned about their companies' long-term point of growth. Let's be clear: growth means increasing in size, such as the growth of a child to an adult. Maturity is the completion of natural development in an animal society. Animals do not grow forever. Animals have their own lives with a beginning and an end, but a company does not. Companies wish to grow forever outside the natural life cycle. This is an important and different point of view between natural organisms and companies.

Development for companies is endless; however, they cannot ignore their environment, which is comprised of human beings, society, nature, and international relations. Simply put, companies should change their concerns regarding their management development. Maturity is a final stage in a company's life. Just think, companies could live forever as long as they have a lively interest in maturity, not just growth of size. The majority of companies will not be mature companies with only size on the agenda.

1.4.2   Estrangement Results Between Welfare and Gross National Product

Results of the Index of Sustainable Economic Welfare (ISEW) in the study by the Stockholm Environment Institute (SEI) in 1996 show clear estrangement between welfare and economic growth and between ISEW and Gross National Product (GNP). In a comparison of 1950 and 1990 growth, the SEI introduces a striking divergence. Applied to the UK, ISEW indicates that ISEW has risen only marginally by 3% despite a 230% increase in per capita GNP; the US is 300% in ISEW and 460% in per capita GNP; Sweden is 250/200%; and the Netherlands is 300/290%. According to the International Institute for Management Development (IMD) (IMD 1997, 1998, 1999), results of a survey indicate a high correlation between evaluation regarding the QWL and productivity. We must reconsider the subject of wealth, or welfare.

According to the data, people associated improvement of quality of life with productivity improvement. But there are two categories of improvement: wealth and welfare. Economic growth has certainly provided people with wealth, but the question is whether their level of welfare has been improved as a result of economic growth.

1.4.3   One Crucial Assignment of Productivity

Japanese industry increased as much as 181% from 1975 to 1995. As a result, Japanese manufacturers could occupy dominant positions in the world market. In the meantime, the Japanese production system was touted as a unique and effective method to increase productivity worldwide. Afterwards, the crucial improvement rate has been almost 2× or more for the past decade. But most industrial countries have had experiences of such a vastly superior productivity improvement rates. France enjoyed 1.41× production expansion in 1829; Germany saw 1.43× in 1850; Denmark enjoyed 1.51× in 1870; Sweden saw 2.33× in 1880; and finally, Japan experienced 2.08× in 1885 (Nishikawa 1997).

These statistics highlight the purpose of productivity activity. It should create the positive welfare of a society rather than mark competitiveness in the market or a higher number of products. In this book, note that welfare doesn't just mean private life; it encompasses relations between citizens and society or community, employees and employers, developed and developing countries, and so on. The balancing act happens with the importance of compatibility management of economic domains and social domains. Productivity activities should contribute to not only manufacturing companies' bottom line but also their corporate sustainability.

Current common wisdom in management philosophy may be strictly focused on competition in worldwide markets. According to Alfie Kohn (Kohn 1992), there are two kinds of competitions: structural competition, which is victory or defeat among competitors, and intentional competition, which are the internal targets and/or goals. Productivity should be utilized for the latter rather than structural competition in the way of reinforcement of cost and/or price competitiveness through productivity.

One of the philosophies of lean production is that it is a system. It is actually not that complex. Lean production means removing flab at the shops. But you can imagine that the reason for such flab is management, staff engineers, and product designers rather than employees on the shop floors. The practice does not accept any waste on the shop floors. This type of management puts a huge amount of pressure on employees and suppliers. A more valuable way to motivate employees is through *engineering competitiveness*, or *business areas' competitiveness*. Japanese manufacturers come up against very tough competition with Chinese manufacturers today due to the enormous gap in wage standards. But it is entirely possible for Japanese manufacturers to succeed if they can enforce products and/or process design competitiveness. Cost reduction techniques in the production process are easy to copy as well. Engineering ability should be applied for real competitiveness enforcement rather than technical skills at shops. A company that's eager for cost competitiveness or price competitiveness does not accept even a small amount of waste, so a tough target of productivity is set for competitive domination.

1.4.4 Company Dignity Should Be Enhanced

What is the objective of productivity? It should be enhancing company dignity. Productivity activities should always be intended to strengthen dignity, or ethical practices and treatment of people. It should be mentioned that a few companies are concerned with both economic and noneconomic issues. According to my study of them, it is possible to achieve a higher level of maturity rather than growth with a hierarchy of five categories.

The first step is to set up a corporate culture of these principles when beginning the business. The second step is to establish economic advantages such as sales volume, market share, and profits value. The third step is to address the need for a philosophy that incorporates social contribution. Many companies are interested in social contributions that are not related to their main business. These activities can include involvement with museums, art galleries, or sports teams, for example, or such as eco-projects that help the environment and the community as a whole. The fourth step is having an irreplaceable presence in the community. The highest level of maturity is to be admired by society.

Five aspects of high-level dignity are: humans, products, social relations, amenities of working, and international relations. A highly dignified company can manage a plural system, not just a single system. Their economic target is profitability rather than profit or market share. The end of the era of mass production, mass sales, and mass consumption is imminent. The objective of productivity enhancement is changing the definition of company growth. Compatibility of economic domination and the contribution of social connectivity should be the objective of productivity management (Sakamoto 2002).

1.5 Strategy for Manufacturing

To enhance cost and price competitiveness, management must know how to effectively control factors in the cost of manufacturing, which generally means controlling personnel and related costs. Since this book introduces the methods design concept (MDC) and work measurement, the relevance of management indicators is equally examined. Approaches to production have been proven effective by actual companies, but first their strategy had to match the new roles for personnel, which would change with the improvements.

Let's remember that strategy is the management behavior that meets the changing conditions of corporate climate regarding external and internal matters.

What role do the weaknesses and strengths of manufacturing play? The competitive circumstances of manufacturing are shrinking. There are various strategies for improving corporate performance. There are several management tools but industrial engineering presides over most touch points relative to international competition. There is no need to acknowledge kaizen and participating management in the supporting initiatives that are quite common today in manufacturing, but we should apply greater emphasis on strengthening manufacturing strategy and top management's interest and attitude toward it.

Let's turn attention to Figure 1.7. Utilization of productivity improvement results such as capacity increase and worker redundancy is also an important topic of strategy. Two categories of utilization exist: passive utilization and positive utilization of human resources. Both are effective in improving corporate competitiveness and performance. Short-range strategy in which passive utilization is

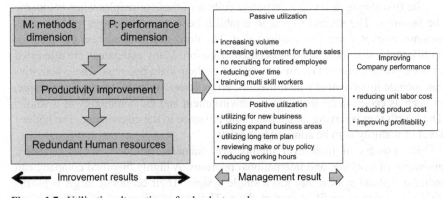

Figure 1.7 Utilization alternatives of redundant employees

applied is satisfactory, but basic or middle-term strategy is required for positive utilization. This is not an easy matter to define and implement company wide; it takes time and effort by employees who must be led by excellent change managers. Otherwise, any productivity improvement activities are just enjoyable for a short period of time like a game that you tire of. When a company implements adequate techniques that guarantee real gains of productivity, manufacturing strategy becomes trustworthy alongside marketing strategy, corporate strategy, and research and development strategy.

Unfortunately, even a few manufacturers believe that manufacturing strategy is a weak practice today. However, as shown below, there are successes.

1.6 Case Studies: Successful Companies in Productivity

1.6.1 Productivity

There are two types of processes for increasing productivity. The first type is buying productivity, which means buying facilities and machines for improving productivity. Buying productivity necessitates a large investment, not a small amount of money (Helmrich 2003). Productivity can be increased by buying productivity without creating ideas for a production method. This is simple to implement, but the productivity level after being implemented is not easy to identify in terms of successful competitiveness. The robotics line of a welding shop in a car manufacturer is a typical example. The second type of process for increasing productivity is creating productivity, or creating ideas for combining ideas from outside sources. The cases introduced in this book speak to the nature of this type of productivity. Other features of creating productivity are less expenditure and knowledge of methods rarely open to others. When these ideas are used by a company, it provides great examples of optimizing human resources.

As shown in Figure 1.8, a producer of frozen foods has had more than 300% productivity improvement. The methods of productivity improvement are reducing the number of workers with MDC and a performance control system using an engineered standard time base. The number of employees decreased from 147 to 97.

In Figure 1.9 company C, the producer of raw materials for foods, cosmetics, and pharmaceuticals, earned a 510% hike in productivity. This company is part of the process industry except for one department of pharmaceutical manufacturing. They concentrated on reducing the manning number for current production capacity with MDC, and then set standard time for all shop floor operations and a performance control system that improved the performance level of workers. When MDC activity started, 164 employees were on payroll; that number reduced to 65. As a result, productivity increased 171% within 2 years. MDC continued to improve production areas, raising production to 252%. It took 1 year to set engi-

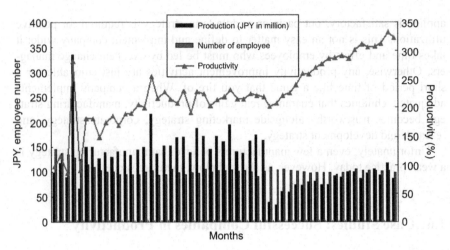

Figure 1.8 Productivity improvement case: company B

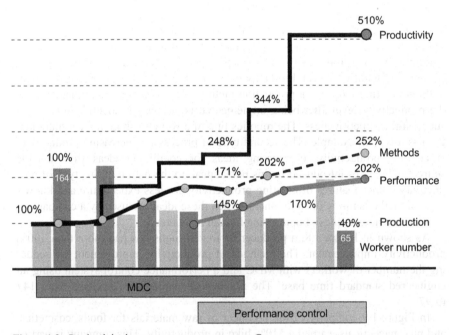

Figure 1.9 Productivity improvement case: company C

neered standard time for all shop floor tasks. Performance control took 2 years to reach 120%, and from the beginning of the activity, the results amounted to 510%.

Other information to note is that with human resource reduction alone, the productivity had risen to 200%. Profit margin improved to nearly 40%. This is an historical record for the company. Lucky circumstances or a thorough improve-

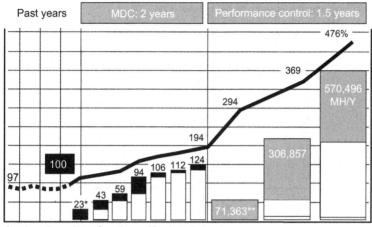

* reduced number of workers **reduced man-hour

Figure 1.10 Productivity improvement: company A

ment in productivity based on hard work? The numbers speak for themselves, but this is a different time globally, and this company may need to consider increasing the number of workers.

Figure 1.6 shows the utilization of 94 redundant areas that management has noted to be weak including: quality control, preventive maintenance, production, and manufacturing engineering. Some of these areas moved to general support staff and sales.

To show a different side to these tasks, Figure 1.10 with company A displays remarkable productivity improvement. The resultant reduced ULC from JPY 8,175 to 3,035, (about 100 JPY = 1 USD). ULC is indeed at a competitive level. Company A is a producer of refrigerators, cubic icemakers, and beer service machines for the restaurant and hospitality industry. They started productivity improvement activity in five manufacturing departments of two plants. Using MDC, they reduced staff by 124 workers within 1 year for a 50% reduction of direct workers and then began performance control for all workers. Additionally, the company improved labor performance from 53 to 130% with engineered standard time; this is 245% productivity improvement. So, all improvements of productivity combined total 476%. The cost value is JPY 3,055,000,000 per year and JPY 12,220,000,000 for 4 years.

However, in this case, as production decreased these results did not remain steady. Redundant employee tasks as a result of productivity improvement are required to be used effectively to maintain company momentum. Remember the difference between passive utilization and positive utilization. The long-term point of view recommends positive utilization for expanding business areas.

Table 1.1 shows productivity improvement in companies who improved M (methods) and P (performance) with MDC and performance control. Their productivity improved 3 or 4× as a total of M and P.

Table 1.1 Productivity improvement with M and P dimensions

Company	Productivity improvement (%)		
	M × P	M	P
A	306	132	232
B	287	177	162
C	401	176	228
D	299	169	177
E	235	178	132
F	490	192	256
G	292	164	178
H	476	194	245

1.6.2   Profitability

Figure 1.11 displays a formula for ROI. The first element is operating profit; it is possible to improve operating profit through cost reduction activity and productivity improvement. The second element is the number of capital turn. How many assets are required to achieve a certain level of sales turnover? Controllable assets in daily management are WIP, products inventory, machines, and facilities. A key management point regarding profitability is increasing profit without increasing assets. There are many alternative methods of cost reduction, but management must be mindful that cost reduction must be achieved without expenditure or investment. Regarding WIP and inventory, about 25% of those values are in the category of inventory maintenance cost; they increase cost and decrease profitability.

The just-in-time (JIT) method is effective in this situation for recovering a company's performance within a very short time as a cost reduction, but not for a matter of productivity. Mechanization and automation would be of interest to management to "buy" more productivity.

Another point about cost is that when making an investment decision, consider benefit vs. expenditure. Expenditure is easy to sum up as the spending of money;

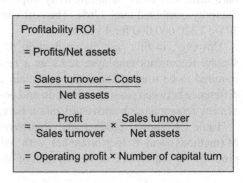

Profitability ROI

= Profits/Net assets

$$= \frac{\text{Sales turnover} - \text{Costs}}{\text{Net assets}}$$

$$= \frac{\text{Profit}}{\text{Sales turnover}} \times \frac{\text{Sales turnover}}{\text{Net assets}}$$

= Operating profit × Number of capital turn

Figure 1.11   Formula of ROI

however, measuring "benefit" is not easy. Again, this is where the concept of "real gain" comes into play. For example, if an improvement has the possibility of reducing 1 hour, does this 1 hour work as a real reduction of a work-hour with payment? It probably would not impact labor cost reduction. A calculated 1 hour is utilized fine as 1 hour of a paid work-hour. This is why a simple expression of benefit vs. expenditure has to be avoided. It is possible to implement engineering economy to approach this type of decision.

Let's consider capital expenditures by first examining Figure 1.12, which shows three components of ROI contribution. There are two effective management decisions involved. One is reducing inventory and/or WIP, which is not too difficult to do. The other is postponing investment opportunity. It means postponing installation time as long as possible. The time value of money has to be reasonably studied. Not only in terms of cash flow, but today's 1 million yen will not be the same value as 1 million yen 5 years from now. The second component of ROI contribution is a cost matter: labor cost with its productivity, inventory, and maintenance costs are reduced. Machines operating costs are also reduced. A third component is utilizing this reduction of cost and productivity improvement for increasing sales turnover. Reducing inventory does not create an opportunity to increase capacity without any expenditure or sales price reduction. Lower price setting will pave the way for more sales, but close consideration must be given to redundant human resources first.

Productivity improvement is not completed as a simple change, but work must be done effectively for profitability touch points; otherwise, productivity improvement will just be a temporary, challenging game.

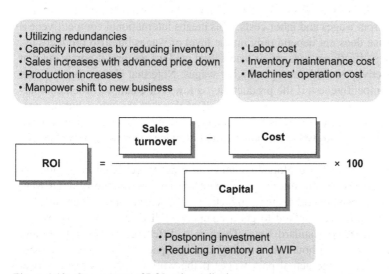

Figure 1.12 Components of ROI and contribution

   Regarding capital investment opportunities, industrial engineers can contribute
to management in decision making, but the steps must all be toward profitability:

- Regarding change we must differentiate between capability and necessity;
  advanced engineers and management who require monetary investment to go
  toward a change initiative insist on capability rather than necessity. But this
  move is dangerous.
- Effectiveness and efficiency must be an ongoing practice. In general, the necessity of capital investment is normally the result of a need for capacity increasing. The reason is to meet market demand or produce new products, for example. Capital investment for capacity increasing is one way to do this, but there
  are other ways more in line with effectiveness and efficiency. It is not unusual
  that management does not investigate the feasibility of effectiveness and efficiency improvement. This should be a real concern company wide.
- Objective summary of expenditures. There is a process that sums up required
  expenditures; however, the calculation should be a neutral and rational one.
- The theory of engineering economy: should be applied. There are a lot of useful
  points in making a reasonable decision.
- Save total expenditure as much as you can.

1.6.3   Effectiveness in Unit Labor Costs

Moving production work to developing economies appeals to management, principally for the benefits it brings in the form of lower labor costs per worker per
hour. But is this a correct decision to make? The relationship between labor costs
and productivity is shown by unit labor costs (ULC). ULC is calculated as the total
paid wages divided by productivity. ULC potentially indicates the contribution to
productivity from wages and labor costs. This means international competitiveness
on cost or price does not depend only on the level of wages but also productivity.
ULC indicates effective wages rather than actual paid wages, and it points out that
higher productivity should equate to higher wages. Note that cheap wages do not
mean cost competitiveness if the productivity is low. For example, when comparing production facilities, if labor costs are half of those of the other, but the productivity is also half, then there is no advantage in terms of cost competitiveness.
Conversely, a plant with twice the labor costs but twice the productivity does not
suffer in terms of cost compositeness.

   It is common to go to developed countries in order to get labor cost competitiveness in the world. From the laborer's point of view, is this the right answer?
No. We have to insist on enhancing productivity that reaches a dominant level of
productivity. A practical method to get a company's ULC is taking those wages
divided by engineered standard time. There is a difference between ULC and
actual wages per 1 man-hour. At working standard pace, those two values are
identical. ULC is $2\times$ the actual paid wages if the working pace is 50%; conse-

quently, the ULC becomes half of actual paid wages if the pace or productivity improves 200%. This pace makes it possible to vary cost, then productivity improvement and labor cost link directly. This makes it possible to understand the flexibility of labor costs and the reserve force of productivity capacity. Figures 1.13 (Sakamoto 2001) and 1.14 show ULC reduction with productivity improvement. These two cases show a higher value of ULC than it was before productivity improvement was more than 2× the actual payment. The ULC becomes cheaper than actual payment after productivity improvement.

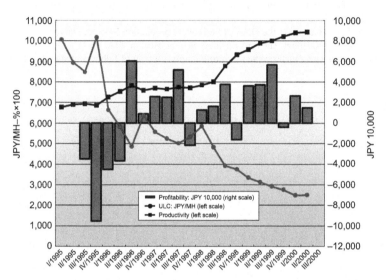

Figure 1.13 Productivity and ULC: company B (Sakamoto 2001)

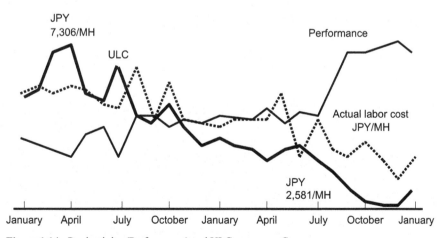

Figure 1.14 Productivity (Performance) and ULC: company C

References

Drucker PF (1973) Management: Tasks, responsibilities, practices. Harper and Row, New York

Helmrich K (2003) Productivity process: Methods and experiences of measuring and improving. International MTM Directorate, Stockholm, Sweden

IMD (1997) (1998) (1999) The world competitiveness yearbook. International Institute for Management Development (IMD), Lausanne, Switzerland

Kohn A (1992) No contest. Houghton Mifflin, Boston, MA

Nishikawa S (1997) 200 years of Japanese economy. Nihonhyouronnsya, Tokyo, Japan

Sakamoto S (2001) A study of company dignity. Toyokeizai Shinposya, Tokyo, Japan

Sakamoto S (2002) Enrich company dignity rather than economic growth. WAPS News of India, pp. 2–4

Spear S, Bowen HK (1999) Decoding the DNA of the Toyota production System. Harvard Business Review, September–October. Harvard Business School, Boston, MA, pp. 97–106

Stein H (1971) The meaning and measurement of productivity. Bureau of Labor Statistics Bulletin 1714

Chapter 2
Systematic Approach
for Manufacturing Strategy

Chapter 2 reviews the seven obvious losses regarding productivity and profitability while performing a diagnosis, or value stream map, of four productivity boosters that ultimately lead to profit. There are different ways to evaluate current conditions regarding productivity and each result makes a different evaluation for the present. Applying different techniques leads to a different outcome for each shop. The purpose of a feasibility study (FS) is to ascertain the possibilities of increasing productivity and the level of competitiveness. Through FS analysis, the most important project areas are established with regard to improvement potential, priority, likelihood of success, and likely resultant effects. The four steps are: recognizing poor productivity levels and conquering them; eradicating old corporate position; preparing an auditing system for productivity; and carrying out manufacturing processes that involve management through and through, then to accomplish world-class manufacturing (WCM).

2.1 Seven Losses Regarding Productivity and Profitability

There are seven losses that management does not recognize and address adequately. Those are:

Loss of lower utilization of hourly base ability compared to standard time. How do you measure and evaluate workers' performance? There is a world standard for working pace. It is very common that working pace differs more than 20% with standards and without standards. As evidenced by appropriate implementation of standard time and workers' performance measurement in Japan, the performance improvement is invariably almost 200% compared to results with no standards or control system.

Loss of utilized areas of factory square footage. How many areas should there be for work areas, machines, passageways, storage, and other uses? It is much more desirable to have areas within distribution designated for work areas rather than for storage. Building costs are normally expensive, so it is effective not only from an investment point of view if redundancy areas are created, and that those areas are utilized well after labor productivity improvement.

Loss of lower utilization of possible operation hours for shift hours. How do you measure the right utilization for shift hours? Is it correct that measurement that shifts hours minus nonworking hours is based on reported idle time? No, because there is 20–30% of difference in utilization between this calculation of results and calculated results of time value on standard time that is theoretical working hours to produce reported products. The difference is between decomposition utilization and piling up utilization. The latter utilization is a more reasonable measure for management.

Loss of lower utilization of machines and/or facilities within a reasonable capacity. This is quite similar to the point of view above, but the number of shifts in conjunction with utilization machines full time without break time (even operators) includes set-up/change-over time, a short break for lunch, coffee breaks, *etc.* The theoretical maximum utilization for 24 h is 100%.

Loss of difference between production and shipment. How about the relative value of production and shipment? It is easy to measure if you can write up "accumulated charts", which equal the accumulation of production and shipment on the Y line and production days on the X line. The vertical difference in production and shipment is indicated in inventory level and horizontal indicates production lead time.

Loss of inappropriate investments. This is the profit margin spent for investment, allotment for shareholders, and employees. Innovation of products and/or processes is developed through adequate investment. Utilizing old machines or facilities that are depreciated may look like cost-saving measures, but this move might result in lost processing time. Innovation speed is rapid today. Developing concrete planning of ideal "dream factories" is possible with an established investment for new machines or facilities.

Loss of management does not improve profitability. Do you measure profitability every month? Is the outcome a key measure or index for corporate performance? This is the most important aspect of losses that is often not recognized by top and middle management. Nothing happens without management's interest or measurement.

These losses are not simply waste. Because management does not understand and implement current practices, nobody realizes the losses they have. Unknown issues are not identified; therefore, they cannot be improved upon.

2.2 Feasibility Study of Productivity Improvement

2.2.1 Difference Diagnosis and Different Results

Let's look at Figure 2.1 (Sakamoto 1991) which introduces different ways to evaluate current conditions regarding productivity, and shows that each result makes

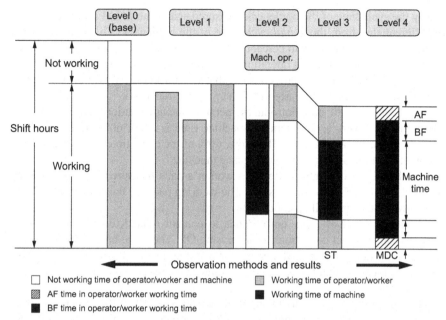

Figure 2.1 Study methods and found results (Sakamoto 1991)

a different evaluation for the present. Note that applying different techniques results in a different conclusion for a given shop. There are 5 levels (0 to 4), with the upper level being more reasonable. For example, you find more than 80% of operators' working time for the shift through work sampling (WS) study. "0 level" is the base of this explanation, but it is a common conclusion that industrial engineers use to identify the current condition. WS study as a tool is applied for evaluating the current condition (0 level). WS study is done by dividing working or nonworking time in a shift; the conclusion will be 80% of working time in one shift. This is why 80% utilization is not bad or good; there are no special problems in the shop, so the rate is average.

The first level of study brings in line balancing. Line balancing is normally approximately 90%; 10% of balancing loss is shown. What does 80% of WS results mean? This rate of working-hour utilization is correct only for the bottleneck worker. Any other workers who belong to the line have 10% balancing loss as standard work contents distribution by industrial engineers. This means that workers at nonbottleneck stations do not work 80% on line-balanced work contents, and this line balancing loss is impossible to find with WS at shops. Balancing loss value is not easy to recognize at shops because the pace of workers is adjusted to meet bottleneck pace by each worker. Besides, it is easy to find small idle or waiting time through analyzed multioperator charts such as a man-machine chart.

Workers often wait during the processing time for completion of machines. If the cycle time of the bottleneck is not set as an engineered time standard such as MTM, nobody can recognize the validity of the bottleneck or standard time. That is why the cycle time of the bottleneck measured against an MTM standard should be differentiated from the *actual* cycle time of the bottleneck. This cultivates the meaning of working time ratio percent in terms of movement of hands and/or body elements, or physically doing operations. An important point of recognition for the operation or body elements movement is that it contributes directly to increasing output (OP), but that is not waste just for supporting functions.

The second level of study introduces operator-machine relation analysis. Taking punch press workers as an example, machine time is about 60%, so the worker operates in the remaining 40%. Therefore, 60% of the cycle time is not real working time; it is time processed in waiting on a machine.

The third level of study relates to a worker's pace measured by a worldwide standard such as MTM or rating scale. The working pace for the worker who does not know its standard will be about 80%.

The fourth level of study corresponds to content related to OP, such as the working contents that contributes to increasing OP. These are basic function (BF) and auxiliary function (AF). BF percent for working time is almost 60% with 40% left of AF, meaning that the work does not increase OP (BF and AF are described in Chapter 6).

To summarize these results in an equation, (80% × 90% × 40% × 80% × 40% = 14%). The example shows only about 14% as a worker's real working time for work that increases OP (14% of utilization of shift hours for meaningful work contents).

It is important to take action at shops for current methods, but those are deemed performance dimension of theoretical productivity analysis. Those improvements of performance are made possible by FM and workers themselves. On the other hand, development of ideal methods concerning manufacturing systems, operation methods, layouts, jigs, and fixtures are much more significant for increasing BF. Designing ideal methods that increase BF in cycle time is rarely the job of industrial engineers.

This basic example explains that completely different results can be offered with study methods as simple as the process of utilization. As you understand, a problem or possibility of improvement cannot show itself through simple study like WS. Formal engineering approaches present better possibilities. WS makes it easy to study shop conditions from the perspective of improvement, but the result does not give a reasonable subject to be improved in the first place. Reasonable improvement subjects in shops are possible through engineering steps like detail analysis of method engineering and measurement such as MTM. And again, methods design concept (MDC) gives more room for productivity improvement with a ratio of BF in working time.

BF is approximately 60% of working time before working methods are designed well. This lead 23% × 60% = 14%, only about 15% of working time contributes to increasing production OP, which means that the remaining 85% of working time is potentially subject to improvement. Instead of focusing on current methods in a shop alone, tap into a theoretical way of thinking.

In the conclusion of this FS, Klaus Helmrich summarizes my supporting experiences above as:

The purpose of an FS is to ascertain the possibilities of increasing productivity, and thus competitiveness. Through analysis, the most important project areas are established with regard to improvement potential, priority, likelihood of success, and likely resultant effects. For your company's implementation, an should include the following steps:

Step 1 Select the processes to be studied. Design models for each process category. Establish improvement goals, such as increased capacity or increased productivity.

Step 2 Identify current losses. Study machine and human utilization and mechanical stoppage times. Perform MTM analyses for manual activities. Perform video and sampling studies. Identify balancing losses and prepare production statistics concerning variations in speed, OP, and scrap.

Step 3 Search for possible directions for improvements.

Step 4 Establish the potential for improvement. Calculate the optimum results.

Step 5 Summarize the results.

Step 6 Choose project areas and set realistic targets. Establish time schedules for implementation (Helmrich 2003).

2.2.2 Symptoms and Background

FS are required for setting quantifiable targets that can be met over a period of a few years. An FS should be used for a long-term improvement program.

Now, consider where you visit when you feel ill. There are three options for your care: drug stores, clinics, and hospitals. A person goes to a drug store just to buy medicine, which he knows will improve his body's condition; he knows the effectiveness of a medicine through a commercial on TV, pharmaceutical advice or other. He is responsible for his decision to purchase that medicine. It is a simple way to alleviate the sickness, and the required time may be less than 10 min for the medicine to work.

The person visits a clinic to see a specialty doctor. The clinic doctor asks for a briefing on the person's symptoms and then measures blood pressure, takes temperature, and may check other activity (pulse, dilation of pupils, *etc.*). Then the doctor completes a medical evaluation and records the information. The person receives a particular medicine following the evaluation. The doctor's diagnosis process may be less than 10 min. That is the complete process if the patient becomes better. If the person still feels ill, he takes the next logical step.

The sick person visits a hospital. The person isn't seen by the doctor immediately upon arrival even if he has a temperature, feels sick and wants to see the doctor right away. First there are a few checks and measures: weight, height, urinalysis, blood pressure, and so on. It takes more than half an hour. Then the person waits another half an hour. After 1 hour or more, a doctor repeats steps that were done at the clinic (talking about the person's symptoms, measuring blood pressure and temperature, *etc.*). This takes 5 more minutes.

Why does the hospital go through such a careful diagnosis process? The reason is that the hospital must see the feasibility of symptoms that the patient either ignored until the sickness escalated or which were prescribed the wrong treatment. There are several scenarios that must be eliminated for the care of the patient. The three options for care, even though they took a long time for the patient to explore, collectively resemble a FS of an everyday challenge (state of health).

2.2.3   Points of Feasibility Study Practice

2.2.3.1   Objective Diagnosis

The importance of an FS is to develop top management's interest concerning productivity and/or profitability improvement. An FS can show attractive possibilities based on numerical or concrete tasks to accomplish and set reasonable quantifiable targets.

Let's return to productivity matters. Which option of the decision process would you have chosen for the best outcome? The options make different conclusions and targets, require different timeframes and incorporate a unique vision respectively for diagnosis. An FS for productivity and/or profitability is recommended to follow the way a hospital executes diagnoses. An FS includes an objective diagnosis and quantitative analysis based on theory (extensive diagnosis).

In objective diagnosis, there are actual professionals of productivity who come to shops and point out the possibility of improvement from place to place in a shop. Those professionals are similar to pharmacists who advise walk-up consumers. The shop that is advised can improve current conditions *if* the professional is an expert in that shop's respective field. But how long can workers accept and tolerate such scrutiny? How long will the improvement continue after the professional is gone?

This approach does not deny results, but it is a subjective approach. Targets must be established and this usually happens with a long-term plan. Kaizen activities directly result in productivity improvement, but not only with participative management; a company as a whole must believe that it is a key matter for profitability and productivity. So, the diagnosis should be kept on an objective study basis. An objective basis of diagnosis dwells upon *object areas*. For further clarification, what is the definition of Muda (waste, or something that does not add value) in kaizen activities? It is not a single way that is considered. One person uses the term Muda but others do not refer to it for particular shop floor practices.

This is a typical subjective way of evaluating Muda. Let's consider a particular operation. First, think about production line operation: it is the line balancing of static line balancing (SLB) or dynamic line balancing (DLB). The ratio can be calculated with engineered standard time or measuring. A bottleneck station in

SLB may coexist with another station that may be a bottleneck station in DLB. The actual condition is a single fact, but objective assessment is different and should assume actions that are also different with objective diagnosis as current assessment.

In order to design a fully mechanized production line (for example, camera production), an effective production engineer would probably use SLB. While its line balancing is adequate with SLB, it does not protect from stoppage of the machine, or CHOKOTEI (a maintenance concept used frequently in Japanese manufacturing, meaning the production is stopped due to defects, parts shortage, *etc.*). The feeding cycle of parts cannot be set with a fixed, constant time value. The use of SLB is satisfactory, but DLB might lose time with unexpected circumstances. Designing cycle time balancing for all workstations and float sizes between particular work stations has to be considered to stave off the point of entry of CHOKOTEI. This definition of balancing ideas would be part of an industrial engineer's skill set. Considerations of DLB must be accounted for so as not to make expensive mistakes.

2.2.3.2 Quantitative Analysis Based on Theory

Many steps in a qualitative analysis help to avoid subjective diagnosis. How much loss is there? How much improvement is possible? Can capacity increase with a reduction in man-hours? A plan can be established with a quantitative prospect, such as the type of development organization (project or nonproject), number of project members and their expected backgrounds and the number of years for developing the plan. A project target should be set at the beginning of project activity. There are endless activities like Kaizen and quality control circle (QCC) that can help carry out the plan; however, company-wide improvements can only be made with results-oriented activity; there must be a limited term for project activities. A key reason that management makes this decision along with the prospected improvement is the result of FS for productivity/profitability with a quantitative base.

2.2.3.3 Extensive Diagnosis

Extensive, overall diagnosis regarding productivity and profitability has to be executed fully. Productivity is divided into method (M), performance (P), and utilization (U). These three aspects create a synergistic product.

For example, if top management asked middle management if 200% productivity improvement (2×) is possible, it is difficult to reply "Yes". Why would management set such a target? A productivity improvement increase of 2× is possible through 30% ($1.3 \times 1.3 \times 1.3 = 2.2$) for each of three dimensions, M, P, and U. How much in total cost can be accepted for the target? This is an important component of engineering economy.

Incremental Cost

It is incremental in that the additional cost that will be incurred as the result of increasing the OP one more unit. Conversely, it can be defined as the cost that will not be incurred if the OP is reduced one unit. More technically, it is the variation of OP resulting from a unit change in input (IP). It is known as the marginal cost. Further, according to *Industrial Engineering Terminology* (American National Standard, 1983), there are two types of additional cost:

Opportunity cost: the cost of not being able to invest in an alternative, due to limited resources being applied to another "approved" alternative, and thus not being available for investment in other income-producing alternatives.

Sunk cost: (a) a cost, already paid, that is not relevant to the decision concerning the future that is being made, *i.e.*, capital already invested that for some reason cannot be retrieved; (b) a past cost which has no relevance with respect to future receipts and disbursements of facility undergoing an engineering economy study. The concept implies that since a past outlay is the same regardless of the alternative selected, it should not influence the choice between alternatives.

An example follows: JPY 10,000,000 (approximately 100 JPY = 1 USD) for decreasing five workers, that is, JPY 2,000,000 per worker. This cost of 2M JPY cannot be used to decide this improvement because it is missing the aspect of incremental cost. Again, incremental cost depends on the relationship between increasing cost and improving effect. Total cost is 10M JPY, but the incremental cost for the first worker is not equal to the average 2M JPY, the second as well, and so on. In my experience, reducing the first worker incurs almost no cost, but the fifth one may require 10M JPY or more, because increasing the number of reduced workers from four to five requires a more machine-oriented or automated facility. The result could be that the first is 0 cost and the fifth is 10M JPY. Considering this example, decide to adopt whatever level of reduction of workers would be evaluated as a reasonable cost for not only the object area but also an acceptable cost for improvement areas. This is also a key aspect of diagnosis in general. This is simple to understand in definition, but often missing in practice. To reduce manning from 5 to 4 and 1 to 0 are quite different as you can imagine from an expenditure point of view, because 1 to 0 means eliminating the operation purpose or full mechanization.

Tools for FS are mainly WS studies recognizing: BF ratio regarding M dimension, direct time study, methods-time measurement, analysis for measuring and evaluating labor performance (P dimension), an accumulated chart of WIP, and product inventory (U dimension).

2.2.4 Practice of Feasibility Study

Figure 2.2 is an example of a FS regarding M, the methods dimension of productivity at Tilten, Torslandaverken, Volvo Car Corporation in Sweden. Tilten is a fully-automated station with three welding robots in which small components are added to the top and underside of the floor plan. The station was a bottleneck.

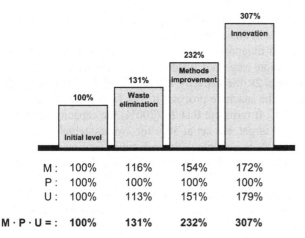

M : 100%   116%   154%   172%
P : 100%   100%   100%   100%
U : 100%   113%   151%   179%

M · P · U = :   100%   131%   232%   307%

Figure 2.2   FS result regarding capacity increase in Volvo

Figure 2.3   Productivity improvement results of FS and actual results (Sakamoto 1991)

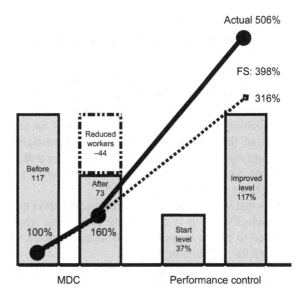

Starting with an index of 100, by eliminating waste it was possible to increase capacity to index 131 (+31%). After method improvements, the capacity was at index 232 (+132%), and after so-called innovative changes capacity was at index 307 (+207%).

Note that there is no need to increase productivity of method innovation for 3–5 years of a short-term plan. This is why work simplification and improvement of changes were decided as a project activity. Therefore, the cycle times of the workstations were reduced by 47% in less than 1 year. In addition, U was increased by 16%. The example shows that there are many opportunities for productivity and capacity improvements, not just in manual work but also in automated process as well. The value of this for the company is gigantic considering the investments that have been made in this type of operation.

Consider another example of diagnosis results at a welding shop. The shop is fully robotic and one workstation was floor welding, which involved a big machine. The diagnosis result was 2× the capacity increase with function analysis of the current waiting time of the machine process. It was found with functional analysis of BF and AF of MDC. It turns out that 2× (200%) the capacity increase was not necessary, so a project target was set at 30% for 2 years, which was sufficient. This means the target 30% is possible with work simplification of method change. This is why expenditure for this change was limited under a small budget.

Figure 2.3 shows another case of FS result and actual result of productivity improvement: about 400% of productivity improvement regarding M and P dimensions of productivity at FS stage. The actual result was 506% of productivity improvement.

2.2.5   Sensitivity Analysis of Profitability

Figure 2.4 is a sensitivity analysis among related subjects regarding return on investment (ROI), otherwise profitability.

It shows possible areas and concrete activity subjects to improve the possibility of profitability in the short term. If you think of the term, *sensitivity* it means acute perception; this is a positive business trait for management to have. As such, sensitivity analysis is a systematic approach that can lead to profitability. The opposite would be random approach, which makes decisions and challenges certain subjects without sensitivity analysis.

Now that we have discussed various upper-level tools in support of profitability and productivity, this is a good time to examine what makes industrial engineers equipped to play a supporting role to management when it comes to capital investment for machines or facilities.

Management is sometimes less skilled in making investment decisions, so industrial engineers can support them in the following ways:

Change initiatives focused from capability to necessity: an investment proposal insists on capability of investment effectiveness. An investment for a machine is justified with the proper forecasting of quality and level of productivity. Industrial engineers ask questions that help map future conditions.

Emphasis on effectiveness and efficiency: effectiveness, such as methods, requires investment. Efficiency in this matter has to be forgotten; the current machine may increase its capacity without methods changing, which would make this a question of efficiency rather than effectiveness. With not only productivity but any subject regarding investment, the two functions effectiveness and efficiency, must be balanced. Effectiveness improvement normally requires higher expenditure but efficiency does not.

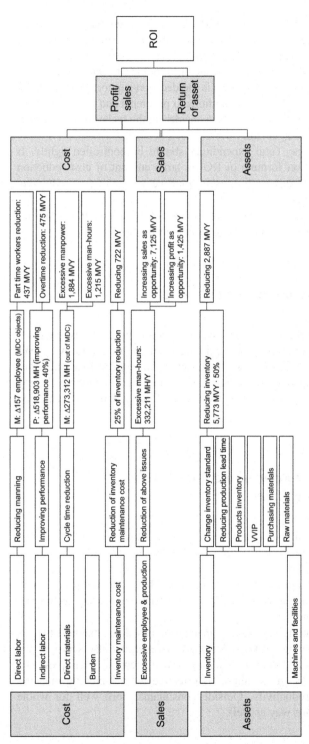

Figure 2.4 Sensitivity analysis of profitability: company C

Objective reports of expenditures: A proposal has to be a summary of expenditure and its effects. A question is objective, not subjective calculation and the items are included or not. Biased, imbalanced calculation is not difficult to find in actual cases, but industrial engineers must be the ones looking for the calculation.

Application of engineering economy: this is the decision support that should be used for all capital investments. It seeks solutions to problems and the economical viability of each potential solution while also considering the technical aspects that can be missed by general management.

Total expenditure savings: Total expenditure should be controlled tightly. If you remember the calculation formula of ROI, a large amount of cost reduction with expensive investment leads profit margin improvement with cost reduction, but return on assets decreases.

In sum, a new kind of management behavior should be implemented, and industrial engineering provides the foundation for this behavior.

2.3 Four Levels of Manufacturing Strategy

There is a different initial position when a company starts productivity improvement as a manufacturing strategy. The beginning must be basic and every employee must understand the steps that will be enacted. The typical four categories of activity are: recognizing poor levels and conquering them, changing the old corporate position, preparing an auditing system for production, and achieving WCM.

2.3.1 The First Level: Recognizing Poor Levels and Conquering Them

The first step of creating a manufacturing strategy is to know the current level of productivity before planning for the future state. Understand the information and activities that are keeping your company from being competitive. For example, a company that possesses engineered time standard will have a higher productivity level than those without such a standard. Basic industrial engineering techniques conquer this level of weakness.

At this stage, implementing scientific management based on standard time is effective in not only improving workers' performance level, but also serves as the objective for any planning and control. Training in management techniques such as IE techniques of analysis and syntheses is effective for heightening management skills concerning productivity. Outside consultants work well for whole organization hierarchy in this way as well.

The following issues and activities are useful once you understand the current level of productivity:

- Losses should be recognized as waste.
- Implement scientific management.
- Adopt engineered work measurement.
- Implement a system that shows effectiveness of management activity.
- Develop a system of quick response for top management decisions to be carried out.
- Cross-train corresponding management.

2.3.2 The Second Level: Eradicate Old Corporate Position

Measuring profitability every month is necessary after middle management takes proper action. These types of managers know cost and cost reduction well, but tend to ignore profitability measurements that reveal results, which often show high-priority problems to solve. Through step-by-step activities, fundamental actions can turn into improvements that outshine competition; for example: level of workers' performance, annual rate of productivity, productivity of direct and indirect departments, productivity improvement subjects such as M, P, and U, production and ordering lead time, and quality.

To turn issues into productive activities, implement a profitability management system, expanding step 1 into an experience for the whole company, identify an outside level of productivity/profitability as a benchmark, and improve the effectiveness of capital investment or improvement expenditures.

2.3.3 The Third Level: Preparing an Auditing System for Production

Preparing a system of objective measurement and evaluation for productivity and profitability is a required project. There are many different cases of successful corporate performance improvement to be inspired by. In other cases, the results are satisfactory, but the approach was a common random approach; management just found problems or improvement objects by chance. Systematic approaches should be adopted for the best practices company wide. Additionally, manufacturing strategy consulting firms have experiences and know how to develop the appropriate auditing system. As a result of auditing, a company can set a middle- and long-range management plan with a reasonable scenario that will guide manufacturing contribution under a corporate strategy. That strategy might be to increase reliability of production departments, give top management consensus for investment, and obtain a long-range view of manufacturing. Outside references as benchmarks are useful for establishing challenging targets for worldwide competition. Reinforcement of top management's involvement in manufacturing divisions will help to define a long-range outlook of manufacturing vision.

2.3.4 The Fourth Level: Accomplishing WCM

The final step is to strive to be the world's "best company of best companies". Pursue a strategy with production as the foundation. Mine the untapped ability of workers for a new production system with the production department itself and/or with an enjoined engineering department. Develop human resources who can strengthen their own skills in daily activities. Enforce the production department's point of view related to marketing efforts and R/D activities. Increase production engineering. Prepare a program not just for short-range requirements such as cost reduction for domination price competitiveness, but with long-range parameters improving the effectiveness of capital investment. Finally, expect a more practical contribution from corporate level of performance.

Staying at the current level using these four steps is simple. Reaching for a world-class level using these four steps is ideal. Planning for a 10-year span of decision making, which is a typical timeframe for the groundwork, demands that top management be eager to improve and have strong intentions achieve a level of WCM. Again, I emphasize that WCM cannot be established in a few years.

Now we look at evidence. Table 2.1 shows practice of a plan in place by company A. This company organized 20 engineers for a productivity and profitability improvement project. Their backgrounds were not primarily in industrial engineering, but rather mechanical and electrical engineering having received training from a graduate university or technical high school. They eventually trained in basic

Table 2.1 A plan for WCM: company A

Step 1	Step 2	Step 3	Step 4
Recognize poor level	Eradicate old corporate position	Auditing	Accomplish WCM
1st–2nd year	3rd–4th year	5th–7th year	8th–10th year
Productivity and profitability improvement	Measurement based management	Compare to benchmark of others	Searching and improvement of long term subjects
Direct: M&P improvement	MBM practices		
U improvement indirect: MOP MBM system	Profitability improvement practices		

courses of industrial engineering including MDC, and Methods-time measurement (MTM). The project the 20 engineers executed covered the three productivity dimensions: M, P, and U. The project activities not only improved productivity and profitability, but also trained the members to be internal consultants. The company, considered the "mother plant", has overseas plants that are now supported by these internal consultants.

References

American National Standards Institute (1983) Industrial engineering terminology. Wiley-Interscience, Hoboken, NJ

Helmrich K (2003) Productivity process: Methods and experiences of measuring and improving. International MTM Directorate, Stockholm, Sweden

Sakamoto S (1991) The MDC training manual. Productivity Partner Inc, Nara, Japan

Chapter 3
General Meaning of Engineering As It Relates to Management

Chapter 3 breaks down the meaning of engineering as it should relate to managers who have been trained in finance and business processes. Engineering is a systematic and theoretical approach for developing higher levels of productivity with the possibility of reproducing the same results anytime. There are several management tools, but industrial engineering is indeed the most useful and promising aspect of engineering. Since participative management has been emphasized as an unexpected contribution for productivity matters, there still seems to be a misunderstanding regarding primary approaches and efforts of industrial engineering. Human resources departments are especially equipped to embody the principles of industrial engineering for the purpose of honing talented teams. Work measurement is defined, as well as the standards of productivity and standard time.

3.1 Definition of Engineering

According to *The Concise Oxford Dictionary*, *science* and *engineering* are described as:

> Science: 1.a branch of knowledge conducted on objective principles involving the systematized observation of an experiment with phenomena, esp. concerned with the material and functions of the physical universe. 2. Systematic and formulated knowledge, esp. of a specified type or on a specified subject. 3. An organized body of knowledge on a subject. 4. Skillful technique rather than strength or natural ability.
> Engineering: the application of science to design, building, and use of machines.

I would personally define engineering approach regarding productivity matters with a more simple meaning: Engineering is a systematic approach for developing higher levels of productivity with the possibility of reproducing the same results anytime. This is the reason, whether a non-expert takes an engineering approach or the task is directed by a skillful engineer, that there is no need for a "guru" or a special expert. Even the young and/or inexperienced engineers that take an engineering approach find a better solution. I can also say that the theoretical approach explained in this book also leads one to find new and unique solutions.

How does your practice fare in productivity matters, especially in methods improvement?

3.2 Management and Management Engineering

Competition in manufacturing is becoming tighter and tighter. There are various strategies for improving corporate performance. There are several management tools, but industrial engineering is indeed the most useful and promising aspect of engineering. An important point to think about when it comes to manufacturing competitiveness and international competition is that there is no need to stress kaizen any longer and urge management to participate. Kaizen is common in today's manufacturing, but you should focus on strengthening manufacturing strategy with getting managers excited about the tools that take ordinary management principles to the same level as that management engineering.

Management engineering is useful in developing a higher productivity level with reasonable costs and developing products that have competitiveness. This is what industrial engineering is all about. Note that management engineering never replaces timeless management skills and principles, but it works well to support management who are searching for competitive methods. These techniques are instilled in industrial engineers.

Webster's International Dictionary defines "management" as the act or art of managing (more or less skilled handling of something). Then "manage" means to train or handle (a horse) in graceful or studied action or stance to control and direct; to make and keep; to treat with care. Within management engineering, industrial engineering is the application of engineering principles and training and the techniques of scientific management to the maintenance of a high level of productivity with optimum effort.

Management cannot manage well with just engineering approaches, but effective tools for improving corporate performance are management engineering. Management is considered an art. This does not mean its practice is simple. Corporate performance is not measured by theories. We should understand the difference or gap between management results and management engineering. There is a lot of literature that introduces success stories from both management engineering and typical management styles. There are many examples of improved corporate performance without effective activities in cost reduction. Achieving results of management is an art in itself. Art is art; meaning it depends on experiences and the nature of the people who are managers. There is no need to stress the building blocks of management with theoretical background alone. The important point is that every company thinks about effectiveness of management engineering such as industrial engineering. Engineering leads objective and effective programs, and there is a low risk of failure yet a high rate of results with an industrial engineer directing the initiative.

Management engineering for productivity improvement can be taught with the goal of remarkable improvement of corporate performance. The following points are important to know before applying the approaches.

3.2.1 Management Should Always Include Measurement

What measurement of productivity is useful? Results of productivity should be billed as OP or consumed resources as IP. Numerical results include: produced sales value, number of pieces, size of square measure, weight, *etc.* It is desirable that OP is not affected by materials, for example. Sales value is not a reasonable OP measurement for the production division because it varies by products mix within sales results. Products mix itself is not always managed well in the production division, and this is why. The comparison between past and present is also an important measure. This means that the number of pieces, size of square measure, and weight are not sufficient measures for productivity. After all, these items do not indicate effort; they are given factors.

So, what measure is reasonable? Universal OP is measured by man-hours; for example, allowed man-hours based on produced OP whether processing of expensive gold or cheap iron materials. How many man-hours of production are consumed? This question points to the necessity of time standard because it is an effective measure for OP. It is called engineered time standard. Applying predetermined time standards (PTS), such as methods-time measurement (MTM), and work-factor (WF), are practical techniques for measuring OP. The denominator of consumed man-hours then becomes simple. It is not easy to get such a number when comparing outside companies, but it is effective in measuring internal corporate productivity.

To discuss or consider productivity without reasonable measure is like navigation without compasses. Again, remember that one essential of engineering is the objective measure of results. To change current methods and measure with reduced cycle time or not does not have much meaning for the company if the change initiative is not objective. Time and numerical factors cannot be disputed; therefore, they are objective factors that are not only useful in improvement methods but an indication of workers' effort to change their P. This is why engineered time standard is an indispensable concept to apply.

Management is measuring results and taking proper actions to meet targets. So, as an equation, Management = measurement + control.

3.2.2   How Much Productivity Improvement Is Expected?

According to results in manufacturing, required productivity improvement is more than 10% per year and more than 15% is recommended if the company wants a dominant level of productivity. The best world-class manufacturers marked more than 15% productivity improvement and in maintaining a level of advanced competitiveness on productivity. It means 2× of productivity improvement for 4 years. Two or three percent or more improvement in a year is possible without a special, aggressive effort for productivity improvement.

Workers' efforts change naturally over time. As stated in the case studies, some examples of companies that organized inner project activity in productivity marked more than 300% within 3 years. Projects with strong leadership that organized special staff (full-time base and consultants) for the initiatives that used a systematic approach and advanced technology tended to do the best.

Looking at the statistics, Taichi Sakaiya insists that Japanese manufacturers are not in a good position currently among world competition due to three parallel policies among Japanese top management: budget restrictions, lack of awareness of previous year's results and lack of insight about competitors. Additionally, there is no policy that challenges management to strive for absolute best practice as their target. Objectivity and a theoretical background in a measuring system of productivity are absolutely necessary.

3.2.3   Methods Improvement Based on Engineering Approach

Work simplification only concerns waste; improvement, however, is not as basic as work simplification. Some types of improvement implement steps aimed at copying others' successful examples such as Kaizen and Toyota production system (TPS). They simply copy others' success though they never catch up and overtake the originals that they followed for many years. Enter innovation, which means searching out ideas that nobody found before, then setting a corporate target for productivity around these new ideas or strategies. According to authors Curtis R. Carlson and William Wilmot in their book, *Innovation: The Five Disciplines for Creating What Customers Want*, the general definition of innovation is: "Innovations require a synthesis of many ideas to succeed, including the new products or services, enabling technologies or capabilities, barriers to entry from competitors, a compelling business model, and essential partnerships" (Carlson and Wilmot 2006).

Innovation can't happen when creating ideas based on current working methods; *i.e.*, improvement cannot be heightened with the new ideas using old practices. The approach of searching for something better yet with the same ways never allows the company to reach the world competitive level of productivity. Note a simple example of manufacturing process innovation: the calculator. The product used to be assembly work with bolts and nuts, but today, it is produced in the process industry without bolts and nuts. This is a success case of products and process innovation. There will be critics of this statement, but you can't ever find solutions for real higher level of productivity when you do Kaizen or an approach similar to kaizen. That is to say, productivity strategy should change from focusing on the past to reaching for a reasonable, higher target that embodies the future, using different practices that zone in on both productivity and profitability.

3.3 Industrial Engineering and Productivity

Figure 3.1 shows the definition of industrial engineering in the terminology. Industrial engineering is a category of engineering like any other, but with additional unique qualities, as shown in Figure 3.2. Table 3.1 shows a simple but significant contribution for increasing real profit through labor cost reduction.

Since participative management, such as quality control circle (QCC), has been emphasized as an unexpected contribution for productivity matters, there always seem to be a misunderstanding on primary approaches and efforts of industrial engineering, for instance, the notion that workers must "come to the shop floor first" and any problems will be solved or processes improved inside the shop. I do agree with this view if you want to start making improvements using the current methods in that specific department, but what if something else is required? Another skill set may be found outside of the shop. Another notion is, "Waste elimination is a key to success!" I do not agree with this view because our target should be set at the highest level; this incorporates not only elimination of *current waste* but also *current methods*.

The best way for production may not be found in the shops. Industrial engineers should either create or find the ideal methods within the engineering mindset using their knowledge first rather than quickly finding a solution in the physical shop. Experiences based on the shop floor approach is not a bad solution, if the solution can find the unique and effective results, but not everyone can reach the same results. On the other hand, those using an engineering ap-

Industrial engineering

Concerned with the design, improvement, and installation of integrated systems of people, materials, equipment, and energy. It draws upon specialized social sciences together with principles and methods of engineering analysis and design, to specify, predict, and evaluate the results to be obtained from such systems.

Figure 3.1 Definition of industrial engineering (American National Standards Institute 1983)

Figure 3.2 Other types
of engineering

Other types of engineering
1. System objectives relate to human beings
2. Approach to problems as total system
3. Solving problems based on economic issues
4. Consider not only engineering solution but also social science point of view

Table 3.1 IE contributions for increasing profit. From Institute of Practitioners in Work Study, Organization and Methods (1975)

	Active IE activities	No IE activities
Sales	100	100
Materials costs	30	30
Labor costs	13	17
Indirect expenses	20	22
General overhead	19	19
Before tax profit	18	12
Real profit	9	6

proach may achieve more reliable results without having a long history of experience because they know a way to solve problems through industrial engineering techniques they already had.

Consider the example of cutting materials with a lathe machine. One scenario is a skilled worker who does not know much about cutting theory to achieve the task perfectly. Another scenario is a mechanical engineer who does not have a lot of experience in operating machine tools, but he possesses knowledge of cutting theory, cutting speed, and so on. This simple example shows the difference between an experience-based approach and an engineering-based approach. Both approaches are necessary to find and get a higher level of productivity, but in the end, the engineering-based approach may get better results than the experience-based approach. Choosing different approaches could bring you different results.

Put another way, what is your definition of the moon? This may depend on approaching the meaning with your own eyes, the use of telescopes or other technology, or the rate of touch-down by space shuttles to the moon. Their definitions of the moon are totally different. Like productivity, the type of results possible is dependent on the approach.

There are so many different kinds of industrial engineering tools available in the world. Any of them are useful techniques to apply when taking an engineering approach for productivity matters.

3.4 Necessity of Facts (Work Measurement)

In this section, let's answer two questions: What are the standards of productivity? And why is standard time effective?

There is no engineering without measurement. This statement not only refers to industrial engineering but also to mechanical engineering, chemical engineering, and electrical engineering. This is because there is no objective judgment of productivity improvement if it is not measured against standards.

The three aspects of productivity such as M, P, and U are measured best with standard time.

Methods, or M, is measured before improvement methods divided by the improvement methods standard. For example, standard time before improvement is 12.00 man-hours and 6.00 man-hours for improvement methods are calculated as $12.00/6.00 \times 100 = 200\%$, double the productivity change of methods improvement. Reduction of standard time for a particular operation is the effect of M, methods change.

P is standard time divided by actual working hours; such as $6.00/7.00 \times 100 = 58\%$. Full capacity (100%) of performance level means a worker just followed standard methods precisely, along with standard pace. Nobody can truly evaluate or measure workers" performance without engineering standards.

And the last is U. The U contribution to productivity is not as simple to measure as M and P because U includes production planning and control, facility maintenance, quality matters and other responsibilities that belong to supporting staff services. How much impact can productivity make with production planning? The number of production opportunities and its total set-up hours may actually reveal productivity losses. Having stoppage hours is the wrong outcome for productivity, and quality is lost with this loss of production as well.

References

American National Standards Institute (1983) Industrial engineering terminology. Wiley-Interscience, Hoboken, NJ

Carlson CR, Wilmot WW (2006) Innovation: The five disciplines for creating what customers want. Crown Business, New York

Part II
Theory of Productivity

Chapter 4
Definition of Productivity/Requirements for Improving It

Chapter 4 defines productivity as the quantitative and qualitative results of the IP of all resources. In management terms, the objective must be to keep effective and efficient balance of management results and management resources. Productivity means how much of management resources are required for a certain amount of management results. This is why productivity is required to improve constantly no matter the economic climate. A "systematic approach" is a definite and organized way of improvement by looking at the target based on a wide and high point of view. It is also called a "problems-oriented approach".

4.1 What Is Productivity?

What is productivity? According to the book, *Industrial Engineering Terminology* (American National Standard Institute, 1983), the definition of productivity is "the quantitative and qualitative results of the IP of all resources. The most widely used productivity measure is one-dimensional (one measure of IP and one measure of OP) and defines productivity as OP per labor IP (*e.g.*, number of trees planted per employee hour, *etc.*). A broader and more modern view involves value measures, *e.g.*, labor productivity equals value added per employee. Value added is defined as the net contribution of business to the value of the IP".

In management terms, the objective must be to keep effective and efficient balance of management results and management resources. Productivity in this book means labor productivity. It is possible to measure the productivity level and compare it to others if management is inclined to know. The calculation equation of productivity is simple, but the meaning should be highly regarded (Figure 4.1). Productivity is calculated as the following: productivity = OP/IP = produced management results/consumed management resources.

Figure 4.1 Productivity is measured

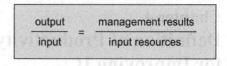

$$\frac{\text{output}}{\text{input}} = \frac{\text{management results}}{\text{input resources}}$$

It is well recognized, however, as simply a ratio of OP divided by IP. Obviously this is not enough clarification or context for defining productivity in the management field. The formula shown is not good enough for defining productivity because this formula is used for any measurement of efficiency such as measuring efficiency of power machineries with gasoline, diesel, and so on, for example. There is also a common misunderstanding between production and productivity. Some believe that production volume increase means productivity improvement. This is also a fundamental misunderstanding of productivity. The productivity level does not depend on the size of the production volume. Productivity is possible with either reduced production volume or increased production volume. At least that is the traditional point of view.

For management's purposes, productivity is defined as consumed management resources, such as time, number of workers, materials, money, and energy for producing management results; results can include sales volume and production volume. Also, results are not dependent on prosperity or depression in economic circumstances. Productivity means how much of management resources are required for a certain amount of management results. This is why productivity is required to improve constantly no matter the economic climate. Changing production or sales volume and productivity improvement are not the same parameters for a company's success. Note that different management actions are required for each of these internal activities.

Production and *productivity* are totally different subjects. That is to say that productivity does not improve simply with increased production and *vice versa*. Productivity is simply a ratio of IP and OP items. Unfortunately, management often mistakes productivity improvement for increased production and their respective values. The productivity level is not dependent on the size of production volume.

As you can understand now, it is possible to increase productivity even with reducing conditions and opportunities for production. Productivity improvement does not require OP increases. Productivity should be a top interest of management at all times. However, in actuality, the subject of productivity is bleak at the management level and this has been a phenomenon throughout my career. Management has a reputation for being interested in short-term actions around productivity when OP increases due to economic circumstances.

Productivity is a result of management regarding IP resources and results as OP. High level or low level productivity by workers is absolutely dependent on management activity that supports the infrastructure for it. This is totally up to the quality of management.

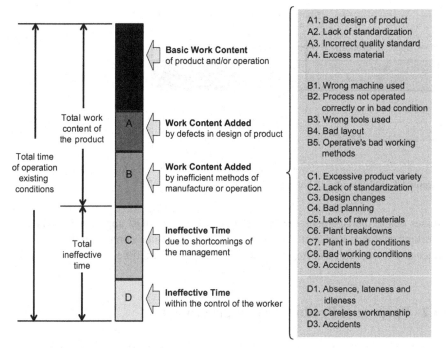

Figure 4.2 Contents of total time (From ILO 1992. Sakamoto rewrote original with ILO permission)

The book, *Introduction to Work Study* (ILO 1997) ultimately divides the contents of manufacturing time into 5 areas. It starts with 2 headings, "Total Work Content" and "Total Ineffective Time". First, the heading "Total Work Content" is divided into 3 areas: "Basic Work Content", "Work Content Added by Defects in Design or Specification of Product" (A), and "Work Content Added by Inefficient Methods of Manufacture or Operation" (B). Then, the second heading "Total Ineffective Time" is divided into 2 areas: "Ineffective Time Due to Shortcomings of the Management" (C), and "Ineffective Time Within the Control of the Worker" (D). "Basic work content" is a very significant expression. Any products produced by different companies of basic work content are usually similar when following this point of view. See Figure 4.2.

Actual production time, however, is quite different for each company; it depends on their corporate culture, management style and the morale of the workers.

BF can say that the amount of BF work is the same theoretically for a product design. For example, you see a certain image in your mind when you look at cars, and then you can explain the different features based on what you remembered, but there is probably no big difference among those cars. For example, there are four doors, front and rear wheels, and windows, those are the BF of cars. Figure 4.3 shows the contribution of MDC and performance control in operation as

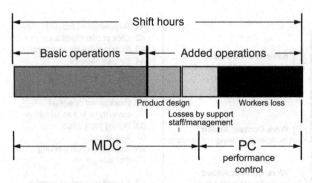

Figure 4.3   Relation of shift hours and MDC/Performance control

losses due to management, support staff, and workers. MDC approaches find new working methods that focus on basic work. Performance control approaches reduce any losses with measurement by engineered standard time.

4.2   Purpose of Productivity Improvement

Higher standard of living. The general meaning of productivity for employees is improving the standard of living by less working hours, higher wages, and more and better jobs, for instance. The number of average working hours has decreased over the last century though production volume has increased steadily over time all over the world. It is easy to understand having less working hours in countries with higher productivity than others.

We can say higher wages have been accepted because of productivity improvement. There are no companies that accept a higher wage level without productivity improvement. The meaning of productivity is not just followed by a number related to production, but also the value added.

Job satisfaction. The balance between employees' physical and mental abilities and requirement for jobs has been much enhanced by productivity improvement. Simple and highly labor-intensive jobs have been improved through mechanization, automation, and ergonomics (human engineering). It is easy to find traditional labor-intensive work in factories with old-fashioned production methods or a general undercurrent of lower productivity. In truth, higher skills are not required for such labor-intensive duties since most of the work day is generally made up of physical activity.

Domination of competition. Global competition is common today not only in prices of labor, materials, and overhead costs, but also in the level of productivity. Moving production work to developing economies is of interest to management, principally for the benefits it brings in the form of lower labor costs per worker per hour. But is this a correct decision to make? The relationship between labor costs and productivity is shown by ULCs. For example, when comparing production facilities, if labor costs at one are half those of the other but productivity is also half, then there is no advantage in terms of cost competitiveness. Conversely, a plant with twice the labor costs but twice the productivity does not suffer in terms of cost competitiveness.

There is a simple calculation of labor cost regarding productivity as ULC, which was explained before. ULC show the growth in compensation relative to that of real OP. These costs are calculated by dividing total labor compensation by real OP. Changes in ULC can be approximated by subtracting the change in productivity from the change in hourly compensation. Labor costs such as wages are two times the ULC if the productivity level is half of the normal level. ULC can reduce of real wages by half if productivity is improved doubled, for example.

Lower consumer prices. You need only compare 20 years ago to now in pricing structure. Due to the same fixed specifications but higher expectations among both employees and consumers, you can sum this up as the result of productivity improvement. Lower consumer prices have made it possible to improve living standards considerably without a higher cost burden. Not only people in developed countries but also those in underdeveloped countries can enjoy a better standard of living because of the consumer products that are produced in higher productivity countries. Many companies move their production plants into low-labor cost areas in underdeveloped countries. Does this make sense? Expansion in itself is not an unacceptable goal, but when you look at the decision in terms of the ULC point of view, low-labor cost areas do not mean low cost on labor charges. It might be even between low labor cost but low productivity and higher labor cost with higher productivity.

Environmental issues. These days, the environment is one of the key management issues. The objective to consider is not only resource energy as a component of products but also excessive WIP and inventory, which can be a waste of utilizing materials. Another issue is ineffectiveness and inefficiency resulting from unnecessary work, such as for WIP and inventory. These items might not be a negligible amount of expenditure spending, but the calculation is not easy. The philosophy of just in time (JIT) contributes to this matter as well. As introduced in a later section, production following shipment is an easy way to recognize this condition.

Effectiveness of capital investment. This is a very important issue that management ignores or has a poor understanding of. All decisions related to productivity that utilize capital investment can impact the company greatly. A capital investment in mechanization or automation of human work tasks alone do not lead to effective improvement of products. Present tasks that workers are doing should be improved before an investment decision is made; otherwise, that decision will not bring satisfying results. These decisions can lead to lowering profitability while improving productivity.

Let's examine the critical points of expenditure and the possibility to save expenditure. After all, management must be able to quantify reasons to invest.

Point 1 Necessary specifications of an investment must be examined, particularly for a planned machine, which is a common investment at manufacturing companies. Engineers who are responsible for particular tasks insist that the new technology is necessary and must be installed immediately.

In the case of a foods material producer, the original proposal of expenditure was JPY 35,000,000; after close examination, this figure was reduced to JPY 15,000,000.

Point 2 A machine's increased capacity right away was a factor in management's decision, but sales forecasting was too long range (more than 5 years in the future) to justify the investment. Forecasting is merely the prospect of something, and the investment may not be too positive or negative, but the more a company has practiced long-term forecasting, the more reliable the forecasted outcome. Time is valuable too, and this aspect should also be examined.

In the case of a medicine producer, an original investment proposal was 22,000 ton per year in the production plan. An investigation found a variation of production capacity fluctuations on that day, so important tasks could be planned for to raise the capacity. Reducing production loss resulted in a 3% capacity increase. Another action was to improve processing time; the initiative resulted in an 8% capacity increase. As a result of the improvements, capacity was increased by 20%. In conclusion, the original plan of investment for two large-scale machines was changed to the addition of one small machine, an add-on to the current machines. Capacity increase was planned at 3,700 tons per month, and the result was 400 under the forecast, or 3,300 tons per month. This is why there was no risk of such a reduction. The first investment plan was JPY 1,500,000,000, and it was reduced to JPY 1,000,000,000.

Point 3 Computers' guarantee limits are coming soon. For example, a vendor recommended her company's version of a software upgrade and replacement of current hardware. As told to the consumer, if there is any trouble in the near future with the computer and they do not have the latest version, the company cannot guarantee that it will work well under the now-outdated version. There were no problems with the computer before so there is reluctance to get the new version. Still, these days, it is sometimes cheaper to buy a new computer than invest in one part or one suite of software because of these constant upgrades that the company invests in. However, we know that replacement timing has to be decided by the owner, not by the vendor's recommendation.

4.3 Different Approaches Lead to Different Results

4.3.1 Input Reduction First

A practical approach to effectively influencing company performance, such as profit, profitability, and productivity, is to start by reducing IP resources. The reason is simple: IP resources are fully manageable resources internally; there are no requirements from outside of the company. Start by increasing OP. Any departments that demand increasing IP resources have lost the purpose of productiv-

Figure 4.4 Effective steps of productivity

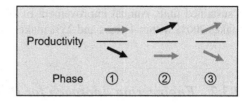

ity improvement in the first place. On the other hand, it is not easy to meet management requirements for an OP increase because it depends on matters outside of the company's control. See Figure 4.4.

The second step is utilizing redundant resources to increase OP results. Utilizing resources is a top management function. These two steps are tools for concentrating on successful productivity improvement activity. After application of these two steps the result would be that the basic level of productivity had increased, and management had directed the balance of IP and OP well. IP reduction can be controlled following a set plan or schedule, but OP increase cannot be controlled, and it takes 1 year or more to get results.

Phase one is decreasing IP resources while maintaining the current level of OP results. This consideration is important for management when they plan productivity activity. What does it mean for your company to target productivity improvement in this way? Phase one focuses entirely on minimizing IP resources for the current OP level. There are plenty of "requirements" given to employees or wishes of a departmental manager to increase OP. However, an important point to consider in the beginning of any productivity matter is to focus internal company policy on reduction of resources. The reason is that the necessary actions are not big issues and the necessary terms to complete this phase do not take a long time. Resource control is a possible control for any company.

Remember reducing working hours, number of workers/operators, and ineffective work tasks become necessary items because of external factors. The beginning of productivity increase should be started at this phase.

Phase two is utilizing redundant resources for current OP requirements. Productivity improvement never advocates firing people; the rationalization is to keep balance between reduced OP and required IP. Closing specific factories is an example of this. To find new business areas of growth, reconsider buying decisions and reducing product prices in order to get more sales orders. On white-collar positions, such as within the engineering department, productivity improvement is possible through allocating engineers to find these new business areas of development.

Phase three is managing two parameters of productivity that depend on companies' requirements. A higher level of balancing between OP conditions and the requirements of IP resources is imperative. How much productivity improvement in a year is a target for companies and why? At least 5% per year is the minimum lowest target, but 1% per month (at least 10% per year) is *recommended*. Excellent companies increase their productivity by more than 20 to 25% per year for over a sustained time. Annual improvement of 10% makes 160% within 5 years, 20% makes 200% within 4 years and 25% makes 200% within 3 years.

4.3.2   Engineering Approach for Productivity

Many management teams have implemented certain methods of improvement with fashionable techniques such as QCC, kanban, and Kaizen. Companies have insisted that top management touted the results of other companies. Even if one way is an effective technique in a company, it is difficult to say the technique is the best way for another company. The decision is usually based on evidence at shops, but again, they may not lead to effective results regarding productivity company wide. It is easy to find failed examples of a technique that was implemented simply because it was fashionable. To define the technique's purpose and proposed reasons for being implemented is the first priority, and then chart the plausible results.

Required by management, the approach involves developing new production methods that can meet the needs of managers who want to run a more lean business. *i.e.*, it refers to measurement to achieve managers' expectations and requirements for cost reductions and productivity improvements, often high targets that are difficult to achieve with the above fashionable approaches.

"Objective approach" leads to recreation of results, or maintains a narrow variation of results. As described earlier, the opposite of objective approach is subjective approach, which offers a lot of differences or a high degree of variation on results.

One subjective approach is a "random approach", which makes improvements one after another with whatever ideas you have. Those ideas in this approach of improvement may be effective at one time, but consistent effectiveness is doubtful and the approach needs further examination. For example, you can pick up one of those universal improvement techniques and implement it without adequate examination of its overall effectiveness. This raises a question about the initiative and capability of the managers. What is the best way to manage problems and issues in our business? Any jobs based on a functional organization should assume an "objective-oriented approach" rather than a "techniques-oriented approach". However, business practices usually depend on a subjective approach to decision making. This book is not declaring that is right or wrong. Generally, it is not an exaggeration to say that there is no perfect way or management style.

To sum up, it is important to stimulate the ability of not a single genius but an organizational genius and to gather different views and opinions of members' organization wide. You must keep a mission in mind when working on your task. You should not just do your best on your job, but also keep your eye on other departments and the company as a whole.

A common approach is to visit a shop and find problems and then improve them from time to time. Some believe that the professional who takes this approach is to be admired, but in this approach the shop floor is not looked at as a system. A "systematic approach" is a definite and organized way of improvement; it looks at the target based on a wide and high point of view. It is also called a "problems-oriented approach". While the random approach can result in a quick solution, the systematic approach seems to take more time.

In other words, the random approach is a realistic and/or subjective way of achieving improvements and finding solutions. It is also called a "technique-oriented approach". Given all these approaches, if you are not an expert on improvement, you should choose reliability over promptness. To put it another way, an approach should be more like a broad plane than a large dot or a thick line. Implementation of the systematic approach can achieve many significant results that are directly linked to corporate performance. Furthermore, the engineering-based approach, MDC, for example, is an effective approach for such purposes as getting a proper headcount in production, capacity increase of major machines and equipment, and setup time reduction. The two approaches for improving working methods were mentioned earlier, but let's recap for a different purpose because a significant issue among management is not taking the time to know the difference in approaches such as "random approach" and "systematic approach". Since there are a lot of tasks in shops that management has to oversee well regarding improving productivity and cost reduction, it helps to clarify the approaches even further.

One approach is technical IE and the other is engineering IE. Within the suite of industrial products, engineers' ideas are more likely to turn into design plans if the engineer has proven process-planning experience. Process design depends on a specific type of management concerned with processes, the design may flourish as a result of exemplary processes. Does this mean any plans are unnecessary?

Engineers who represent those projects with a design target usually possess a background based on one of the fields of theoretical engineering. One such field could be industrial engineering. Industrial engineers design process plans that include production methods, procedures, and the layout of a plant. Engineering thought as a whole is not just for thinking about weak points, defects of products, or defects of the production process. They think about economy and simplified methods, and then set those thoughts into standard operation procedures. Management that supports engineering is expected to be interested in the effectiveness of a systematic approach.

4.3.3 Three Levels of Improvement

Generally speaking, there are three approaches to improving production methods (see Figure 4.5). These are: eliminating waste or work simplification, making improvements, and innovating. The first approach involves removing waste (*muda*) found by observing conditions on the factory floor, as seen in small group

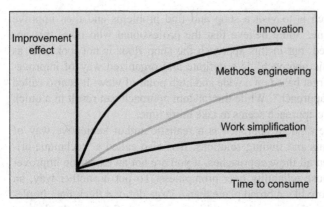

Figure 4.5 Improvement outreach

activity (SGA), Kaizen, and other self-starter activities, and therefore known as work simplification. The second approach entails increasing productivity levels even when there is no waste by taking effective ideas learned from other companies, for example, and directly introducing those techniques and methods. Reductions in retooling and set-up times achieved by many companies through the use of single minutes of set-up are an example of this approach. The third approach involves developing new production methods that can meet the needs of managers who want to run a more lean business, *i.e.*, it refers to measures to achieve managers' expectations and requirements for cost reductions and productivity improvements, often high targets that are difficult to achieve with the first two approaches.

Even if all factory-floor waste (admittedly difficult to define, since what counts as waste may be subjective) is eliminated, it offers no guarantees over discovering new production methods that can give a company competitive advantages. Similarly, even if effective techniques from other companies are successfully introduced, it does not decrease the desire of the company to achieve higher levels of productivity than those of the company that developed them. The development of effective production methods that can meet management needs is something that can be achieved by designing new production methods.

When looking for ways to improve productivity by designing new production methods that will have a direct impact on results, it is important to decide which of the three approaches to take. To explain the differences between the approaches, let us consider a simple example: Imagine a full moon. How would you define it? What I want you to do here is imagine how you would describe what kind of a thing it is to another person.

"Different approaches" could be analogous to looking at it with the naked eye, viewing it through a telescope, or sending a spacecraft to land on the lunar surface. With the naked eye, the moon is romantic and looks like a big ball; through a telescope the many craters come into view; and landing on the moon means that you will be able to see nothing but the desolate lunar surface. In this way, taking different approaches results in completely different views or *definitions* of "full moon". Similarly, taking the three different approaches to production methods can give you vastly different views; *i.e.*, change what you see.

If the management is strongly interested in implementation of all those well-known techniques, they cannot compete with the other good-standing companies who must have gone ahead. Such management may end up as severe management of cost reduction that does not allow even trivial waste. When I say waste elimination, improvement, and innovation, waste elimination refers to a reduction of waste and inconvenience, while improvement refers to a duplicate implementation of competitors' good examples. These are not expected to lead to a comprehensive management approach covering a unique production system and its related matters. Any company requires a strategic management approach that is explicitly conscious of the innovation level from the beginning, no matter what the current level is.

A simple example of improvement and innovation is word processing by personal computer (PC) today. The manual typewriter was *improved* upon: the ink was made darker, the machine itself was made lighter, and the ball element was introduced, *etc.* Word processors, on the other hand, were an *innovation* which had been developed after word processing software for PC. Innovation in personal computer design continues each year.

Three stages are involved in developing more effective operating methods. You can go to almost any shop floor and find some inefficiency. It may be improved immediately in many cases, and implementation is not difficult. Other types of improvement must be developed, however, to achieve the full potential for improvement. They may be classified into the following categories:

- work simplification;
- methods engineering; and
- innovation.

Three methods of change activities are summarized in Figure 4.5. The most significant differences between the three activities are decreased nonworking time and/or AF work or increased BF work. It looks as if they are similar, but real results achieved from each of them are totally different.

4.3.3.1 Work Simplification

This kind of improvement, which is the lowest level of improvement, is normally focused on eliminating unnecessary work or minimizing any inefficiencies involved in an operation based on subjective discrimination that it is waste, such as inefficient work contents. SGAs such as quality circles and the analysis of current methods or processes all belong in this category. While this area provides a great opportunity to achieve results through a large number of varied ideas, only a small percentage of the total improvement potential will be achieved through these techniques. Typically, these improvements do not require major changes in existing machines, facilities, or layouts and consequently do not involve a great deal of capital expenditure.

The first step in work simplification includes suggestion plans, autonomy circles, and waste improvement when it is found in shops by workers themselves. It is effective to share productivity issues between employees (workers) and management. But the results might not have a large affect on corporate performance in general.

Once methods have been improved through work simplification, it may be felt that the resulting working practices should be comparable with other companies or proved management techniques. However, this may not be the case, since varying levels of work simplification potential exist in different situations. In order to match or exceed the productivity of other systems that are being used as a benchmark for acceptable productivity levels, other improvement techniques must be used.

Work simplification is just concerned about waste; improvement is not as limited as work simplification but implements copying others' success examples such as Kaizen and Toyota production system (TPS). Still, they may never catch up and overtake the originals even if they followed their examples for many years.

The methods engineering category will require minor changes to be made to production hardware, for example, improvement to workplace layout, fixtures, machines, and tooling. This often requires a small amount of investment, which may be easily justified.

4.3.3.2 Methods Engineering/Improvement

The second approach is improving current methods through industrial engineering tools with typical or traditional techniques regarding productivity issues. It has been known as taking "scientific steps" for solving problems. The point of this approach is a detailed analysis of current methods in practices. What happens at shops? An inquisitive attitude is necessary for this method as questions may result in finding new solutions for productivity improvement. It might get more effective results than the waste elimination approach, though a question may be too basic for current methods. What is the significance of current methods?

At a competitive level among competitors, how much of improvement is found or not is not an important point. There is a goal that a company should reach for, which is a level better than "competitiveness". Those goals may not be achievable using current methods. In other words, really effective solutions might be out of the scope of certain categories of improvements.

An effective improvement technique developed by Shigeo Shingo was named the single minute exchange die approach (SMED). This approach had a target setup time of less than 10 min. Many companies reduced their setup time at their press shop, for example, to less than 10 min. However, only the first company, who developed the die-setting idea, had a real competitive advantage over the other companies, who were some years behind the original ideas. To make significant progress, therefore, we must develop a new way of thinking concerning methods improvement.

This approach requires time to be spent in analysis of the current situation, to identify problem areas, and to apply common sense to eliminate the waste and poor practices by developing new methods. Traditional industrial engineering training puts the emphasis on the above points, and most industrial engineering training courses and books are about improvement of the current work methods, relying upon small group projects. The advantage of "work simplification" is that the cost and time required to apply these methods are relatively small. The weakness of this approach is that it does not focus on the major opportunities for improving productivity.

4.3.3.3 Innovation

The third approach is a more mainstream concept today and extremely important in today's tough cost competitiveness circumstances. We should leave past methods behind (those that did not have a proven track record) for finding effective solutions that meet future circumstances well. There are three points for practical thinking for reaching innovative levels as much as possible. One is to define working models instead of current methods, consider functions as the basis of methods such as operations or element operations, and then set targets based on objective and reasonable parameters. General approaches to engineering consider this premise. It is easy to understand the difference between improvement and innovation of a product. Let's look at an example:

There are two useful presentation devices: the 35-mm slide projector and the overhead projector (OHP). Using a slide projector is the traditional way of making presentations. Some modern features have been added such as remote control, forward, backward, focusing, and cartridges. These kinds of improvements have made it easy to use the slide projector for speakers at presentations, for example. However, there is still an inconvenience with the slide projector: the main switch. One can turn a slide projector on and off only on its shell and not at the speaker's position. It is very easy to control the master switch at the speaker's position if you install a switch. This kind of improvement is classified as work simplification and/or methods engineering.

On the other hand, there is the OHP. The OHP is easily turned off and on at the speaker's position, and all other points of inconvenience with slide projectors are overcome. The question is whether the OHP was developed as an improvement on the slide projector. The answer is no. The OHP was developed without improving the slide projector. The specification of overhead projection is defined and engineers developed the OHP instead of making an accumulation of continuous small improvements to slide projectors. This is an example of innovation of products and product design.

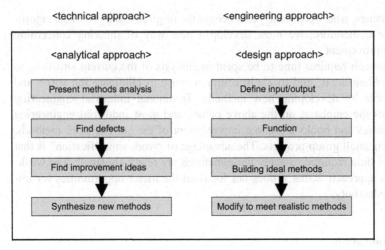

Figure 4.6 Analytical and design approach

Another example of innovation is the air-powered nailing machine, which is fed by a magazine or cartridge of nails. The conventional approach to improving the nailing of wooden structures is to develop improved methods by studying the current operation. The operation is then improved by simplifying the grasping and positioning of a nail and by experimenting with different lengths of hammer handles, hammer head shapes and weights until the optimum combination is obtained. This follows the procedure Taylor followed in his famous shovel experiments. It is the conventional methods engineering approach.

The innovation approach is to study the BFs of the operation, to develop a completely new method of performing them without regard to the current method, and to accomplish the BF in a minimum of time. This is the approach that would lead to the development of a nailing machine to replace a conventional hammer and nail operation.

With previously described types of methods improvement, you may have a large number of very small ideas that can only take you so far. The approach must be changed, or efficient and creative ideas will not be found to raise the potential for efficiency. This level is achieved through the design approach, such as the MDC.

An improvement target must be set as high as possible and by the improvement project itself rather than through the copying of other projects' attainments. The MDC may not be applied to current methods since innovative improvements do not work through analysis of current methods. In place of this approach, suitable methods for defined OP should be created so that innovation results. Having gone through the MDC, innovation will reach the ideal level.

Work simplification and methods improvement are both analytical approaches or technical approach; innovation is a design or engineering approach. When you use traditional methods improvement such as work simplification, you will be

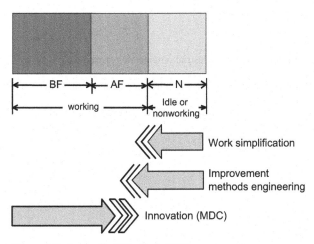

Figure 4.7 Increasing BF ratio is MDC

interested in the distance for getting and putting parts, utilizing both hands, and improvements without changing main parts of jigs and fixtures. If you follow the MDC, however, there is more opportunity to discover a totally different way of improving methods that don't just follow the current, basic operation methods. The point of this difference is whether you concentrate on the BF or not. This is why the MDC will lead to a jump to a new, higher platform. See Figures 4.6 and 4.7.

Within manufacturing, the operations and processes are means of converting from an IP status to an OP status. At a lower level are operational elements and motions, some of which contribute directly to meeting OP demands and others which support prior to upstream work contents. These two categories have been defined as BF and AF.

The most significant differences between the three activities are decreased nonworking time and/or AF work or increased BF work. It looks as if they are similar, but real results achieved from each of them are totally different. The MDC increases BF work for higher productivity improvement and innovative ideas.

"Innovations require a synthesis of many ideas to succeed, including new products or services, enabling technologies or capabilities, barriers to entry from competitors, a compelling business model, and essential partnerships" (Carlson and Wilmot 2006).

There is a simple example of manufacturing process innovation: the calculators. It used be assembly work with bolts and nuts, but today it is produced in the process industry without bolts and nuts. This is a successful case of products and process innovation. You can't ever find solutions of real higher level of productivity with Kaizen. That is to say, productivity strategy should change from comparison to the past in order to reach higher targets, yet this cannot happen without a fully sustainable support system from management. The question is: What productivity target is appropriate for the corporate level of competitiveness?

References

American National Standards Institute (1983) Industrial engineering terminology. Wiley-Interscience, Hoboken, NJ

Carlson CR, Wilmot WW (2006) Innovation: The five disciplines for creating what customers want. Crown Business, New York

ILO (1997) Introduction to work study. International Labour Office, Geneva, Switzerland

Part III
Outline of the Engineering Approach
to Productivity

Chapter 5
Three Dimensions of Productivity

Chapter 5 follows the principle that you follow different ways than you did in the past. Different approaches find different solutions for productivity. How much improvement do you need as the target? Theoretical classification of productivity methodology based on industrial engineering is required. With an effective engineeering approach, it is possible to guarantee the achievement of a reasonable level of results. To reach a higher level of productivity improvement, there are two keys to success. One is to separate productivity contents into M, P, and U, which depend on industrial engineering techniques. Another is to adopt new approaches to each dimension of productivity. M, P, and U encompass total productivity improvement.

5.1 Points of Successful Productivity

Examples of remarkable productivity improvement by Japanese manufacturers reign in the previous figures. The two companies shown are producers of foods and food materials, such as ham, sausages, and powder materials. What they did for the results included improved production methods in shop floors for reducing number of workers with MDC and labor performance control with engineered standard time. They are recognized as the world-wide model for working pace standard.

There are three important points to productivity success. The most important one is "changing approach for productivity matters", which means you can get really different results if you follow different ways than you did in the past. Different approaches find different solutions for productivity. A change in action leads to a change in results. The second point is that you should create your own ways rather than copy others' experiences. You might gain some success through copying but it will likely be short term. Again, you may reach a level similar to the original company's level but the approach does not lead to anything further. Concentrate on creating a solution for what you want, not how to simply make a copy. How much improvement you need as the target cannot be discovered through following fashionable methodologies. The third point is: theoretical classification of productivity depends on objectives and methodology based on industrial engineering. These tools would be the three dimensions of M, P and U. Industrial engineering provides enough experience and background of theory to create lasting results. The engineering way of thinking is more important than just getting results without such a background.

Edward V. Krick said: "Engineering is primarily concerned with application of analytical methods, principles of physical and social sciences, and the creative process, to the problem of converting our new materials and others resources to forms that satisfy the need of mankind. The process involved in solving the conversion problems is ordinarily referred to as design" (Krick 1965).

The engineering approach or way of thinking contains review of obtained results and systematic steps to get those results in the first place. It is not an appropriate approach for those with massive skill or a lot of chances. With an effective engineering approach, it is possible to guarantee the attainment of a reasonable level of results without extensive experience.

5.2 Relationship of M, P, and U to Standard Time

Step by step progression of phases is important for success from the perspective of company performance. There are three possible phases. It is very important that management wishes to reach higher productivity. To reach a higher level of productivity improvement, there are two keys to success. One is to separate productivity contents into M, P and U with dependence on an industrial engineering techniques point of view. Another is to adopt new approaches to each dimension or function of productivity. Those three dimensions multiply productivity results. The M dimension contributes effectiveness and the P dimension contributes efficiency. The U dimension cannot provide clear results without these two dimensions. Synergy of the three dimensions for improvement is ultimately the most effective goal.

Consider a simple example of travel time dependent on traffic methods that have great potential for the future (Table 5.1, Helmrich 2003). For travel of about

Table 5.1 M, P, U on travel time

	Car	Train	Airplane	Potential
Methods	10 h	3 h	50 min	300%
Performance	9–11 h	3–4 h	50–70 min	20–30%
Utilization	Disturbances	Minor delays	Delays normal	20–30%
	To city	To city	Connections	

500 km, a car takes 10 h, a bullet train is 3 h, and an airplane takes 50 min. This comparison is just about hardware such as car, train, and airplane; this is the M dimension of productivity. However, there is a software issue for each travel method as well. Above is the scheduled amount of time, but actual time may fluctuate 9–1 h for the car; the train may be delayed 1 h; the airplane can vary much more; this is the P dimension. Another time loss may happen due to the travel plan; this is the U dimension of productivity. Total travel hours are dependent on these three dimensions. Airplanes means high-cost machines, but it does not guarantee a high level of productivity.

What is interesting about the factors M, P, and U is that together, through multiplication, they give the magnitude of the total productivity improvement. Try calculating the total productivity improvement in the example given below.

Start by calculating the relative improvements of M, P, and U, and then multiply the factors to produce the total productivity improvement.

Following a simple calculation shows the usefulness of the three dimensions' synergy, for example:

$$M \times P \times U = 0.8 \times 0.8 \times 1.0 = 0.64 \text{ case A}$$
$$= 1.2 \times 1.2 \times 1.0 = 1.44 \text{ case B}$$

1.0 means normal 100% and average level compared to the industry average. The U dimension is set at just 1.0 because U does not have a high potential of productivity improvement. Case A is 0.8 for the M and P dimensions; that means a little bit lower than the industry average. It is a condition in which with just a little bit of poor working methods and/or old machines are applied as working methods and a little bit of poor levels of P, the performance level overall is at a lower level compared to the standard. That 0.2 (20%) of lower level is not big enough for either dimension M or P but the result of multiplied M and P leads to 0.64, almost a half level behind for industry average productivity. The results are too far behind the standard, which is not an acceptable level. A question is, do you measure productivity in these three dimensions?

Case B shows 1.44, or a 144% level of productivity compared to the industry average. Only a 20% increase for M, and P results in a 44% increase for total productivity. Generally, 200% (two times) or 400% (four times) improvement of productivity may not be a believable goal, but a 150% improvement for both of M and P leads to a 225% improvement, more than two times of productivity improvement.

5.2.1 Dimension of Methods

M is, without comparison, the most dominant dimension with regard to improving productivity. Well-applied methods also create motivation and better U, since the process becomes more secure (fewer disruptions). It is therefore important to start with methods development when organizing productivity improvement.

Methods can be divided into two categories.

The first one is hardware, such as machines, tools, layout, and the like. Management simply decides to implement new machines or automation ideas for an increase in productivity or to save the number of workers, but it requires a high amount of expenditure. An easy way to apply this productivity improvement starts with the premise that there is no prior competitiveness among competitors because they have not adopted those new machines from outside suppliers. In my experience, I never refused those investments if a higher level of productivity could be proven with this technical innovation. But management has too often taken a mild-mannered stance when dealing with desired OP for their company. Sometimes those ideas to purchase machines are less important than other ideas, such as the proper training of employees.

Anybody travels quickly if they take airplanes rather than trains. There is a tram system called Shinkansen in Japan. The maximum speed is 300 km/h. There have been no accidents since the Shinkansen started business in 1964. Trains leave every 10–15 min from terminal stations. Shinkansen is the most convenient train in Japan, just as certain factories have the most modern facilities of manufacturing. But anybody can take Shinkansen if they like. This means absolute level of travel time is improved but the relative level among travelers is still even; there is no advancement on travel time.

There is a similar kind of misunderstanding made daily regarding productivity issues. Companies believe they have reached a higher level of productivity by implementing modern, high-tech facilities, but no modern facility in and of itself can guarantee a competitive level of productivity.

The second category is software, such as motion patterns, organization, training, and support organization. There is often a deficiency of method on the software side. The danger of limiting improvements to this area is that the lasting effects are limited. As a result, a positive and/or negative effect will be found.

An effect of prolonged focus on the areas where improvements were made is the unwillingness to learn that improvements will not last forever without sustainable management and dimensions of productivity.

Let's take a look at an example of dealing with losses on line balancing through software. Balancing losses can be divided into two categories: SLB and DLB. SLB deals with a careful balancing of the capacity of various stations along the production line by industrial engineers. DLB, which is more dominant lately, is considerably more difficult to handle because of variations in the work content of different products that are manufactured on the assembly line; it is difficult to rebalance the line. Natural quality variations of IP material and workers' performance variation of time in cycles both belong to the DLB effect. One way to deal with dynamic balancing losses is teamwork on the production line, where employees are encouraged to readjust the line by allocating tasks and resources at each work station. In order for the employees to adjust to this flexibility, they must be multiskilled and more eager to take initiative under certain circumstances.

There is often a deficiency of method work on the software side. The danger of limiting improvements to this area is that the lasting effects are limited. An effect is produced so long as a focus is maintained on the areas where improvements have been made. But, the lasting effects do not come automatically.

Changes to the hardware often lead to long-lasting effects once the equipment is in place. A combination of both the hardware and software improvements is the key to success and lasting effects.

It is possible to divide software into three areas: manufacturing system, manufacturing methods, and management system. Another point we must pay attention to is the dimension of M where real competitiveness happens when the other company has advanced software rather than hardware. Hardware domination requires recognizing what has happened with competitors. Meanwhile, software advancement of manufacturing is never open to other competitors. Information related to any internal changes of the software condition should never be disclosed externally, especially to your competitors. This cannot be recognized without visiting and watching their shop floors.

General improvement issues of M are:

- BF time should be the maximum amount of cycle time.
- Movement elements of machines should be parallel to move simultaneously.
- Moving distance of unnecessary movement should be minimized (especially empty movement).
- Work stations' cycle time should be categorized as DLB rather than SLB.
- Set moving speed of machines at maximum possible of logical and reasonable speed.

5.2.2 Dimension of Performance

P, i.e., the motivation of the employees or the speed of the machines, is often emphasized as the means to increase productivity. P is important, not the least since it produces a multiplier effect on the improvement of M and U. It must, however, never be the only measure for improving productivity. Letting employees know how important their work contributions are, combined with clear targets, generates motivation.

The speed of machines often varies greatly. There is a lot of potential in reducing these variations and the possibility of raising the speed to that claimed by the manufacturer. Low machine speeds are often due to technical deficiencies or disruptions. It is only once these have been eliminated that speeds can be raised in a controlled way.

P is the result of measuring and comparing between standard and actual time. Standard time is usually based on the standard methods such as measurement of workers' actual P compared to the standard working methods. Standard time is defined by "given" methods. The most important part of the definition is regarding standard methods. Actual working time depends on the applied actual methods rather than an individual worker's pace. There is a simple misunderstanding about measuring workers' P with working pace or speed, for example. There is a definition of working pace that says standard pace is the definition of standard time. Practices of measuring performance and improving the level of P are primarily due to following standard working methods themselves. Workers ignore the standard methods and supervisors do not understand the necessity of supervising instruction of the standard working methods to each worker at their shops.

Another point is to ignore the global standard of working pace such as task standard, which has been approved as the practice world wide. A practical tool to know this world-wide standard pace is through application of MTM. The measured results where standard time is not applied normally indicate a very low level of P such as around 60% for a standard of 100%. This means an almost 170% improvement potential on performance. Why do you miss the potential without caring about it?

There is no reason for management to disapprove standard time. There is a misunderstanding that management does not like to introduce standard time and/or MTM to their shop floors. Standard time need only be set and shown to workers, along with any supervising actions such as instruction of standard methods.

There are also similar issues concerning both the machines and processing time, because there is no world standard like MTM for them except machine tool operations. However, the difference of standard and actual time is easily found in practices due to a lack of theoretical background regarding current machines and processing time. According to practical experience, the performance level of machines and processing time is a low 80% compared to reasonable and theoretical

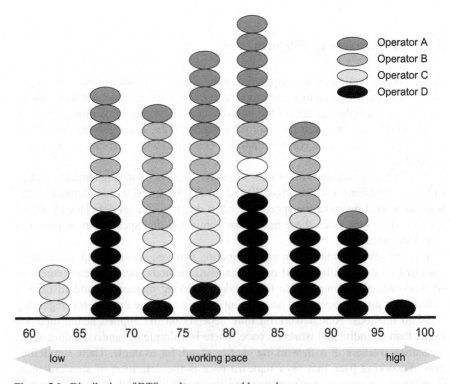

Figure 5.1 Distribution of DTS result on an assembly work

standards of machines and processing time. Compared to a standard of 100%, more than 130% of improvable potential is there regarding the performance level of machines and processing time. None of these figures should be ignored by management although it is small.

The improvable potential of the dimension of P is up to standards itself; industrial engineers are expected to perform at their maximum even at the higher level of standard time. Practice of fluctuation time value is shown in Figure 5.1.

General improvement issues of P are:

- Complete division of job functions among operators and other functional organizations.
- Communicate directly between two shifts.
- Measure P compared to engineered standard time and improvement points.
- The workers should be trained with the best way that is standard working methods.
- Supervisors should stay on their shop floors a maximum amount of shift time to observe and instruct their workers.

5.2.3 Dimension of Utilization

U is an expression, for both machines and humans, of the proportion of planned time that is used for activities and creates value. Common losses for both humans and machines are often caused by technical instability. This can be a result of the poor state of equipment, low material supply, and other various changes in the quality of components. Under an excessive amount of goods flow, balancing losses often occur, because the workload was not equally distributed to all the work stations in the workflow. However, variations in the work content of products largely complicate the matter even further.

In intensive goods flow, balancing losses often arise, such as difficulties in distributing labor equally between all work stations in the flow. Large variations in the work content of products further complicate the matter.

Production planning and control, facility maintenance, and quality control are activities related to the dimension of U in productivity. Those activities regarding U generally are the responsibility of support staff but it is still possible for those contributions to productivity to be measured against reasonable standard time.

U contribution to total productivity does not have a larger potential compared to M and P. However, one thing to remember is that the level of inventory and/or WIP, does not make any impact from the productivity point of view. This means conflict between partial productivity in total plants or company-wide productivity can arise because cycle time reduction only takes place at a specific location, department, or team. That means real contributions to productivity should be increasing the total level of productivity rather than specific shops or work stations. The process must be organization wide. How do we utilize a partial level of productivity to form the contribution of U?

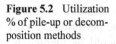

Figure 5.2 Utilization % of pile-up or decomposition methods

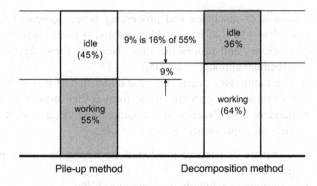

Pile-up method   Decomposition method

There are two interesting approaches regarding U. Those are the "pile-up method" and the "decomposition method" of U. U percentage is normally measured as reported idle or nonworking time subtracted from the base; this means one shift minus reported nonworking time is U. This is the "decomposition method" to measure U: $(100 - 36) = 64\%$. Another approach is to calculate U according to standard time. Total standard time as the amount of finished work is working time; if working pace is 100%, then 55% of U can be achieved using the piling-up method, the remaining 45% is nonworking or idle time.

Figure 5.2 shows a comparison between two methods of measuring U. Working time is 55% for the pile-up method but 64% for the decomposition method. The difference of the two measuring methods is not small; 9% and 9% is 16% of 55%. A difference of more than 20% is not unusual according to practices. This difference is an issue of the U function.

Figure 5.3 is called "accumulated chart". The vertical stem is the quantity of volume regarding production, shipping, and inventory; the horizontal stem is production days. The difference of volume between production A or B and shipment is inventory. The chart makes it easy to recognize over inventory conditions accumulated on a daily basis, such as the difference between a-1 and a-2. This example shows an improvement result for "shipping following production" (similar to JIT). The meaning is production immediately followed by shipment. The situation before improvement was a large gap between production and shipment, such as a high level of inventory. A production plan was decided on by forecasting based on past shipment records; this is the result of estrangement of production and shipment. This means unnecessary production was planned and made. Unnecessary work was done. Another point is production lead time, such as b-1 and b-2. Changing production A to B causes improvement not only by reducing inventory but also by shortening lead time. Figure 5.4 shows the practice of improvement results through accumulated.

Production A is before and production B is after improvement of shipping following production. The improvement results changed inventory level from a-1 to a-2, and production lead time from b-1 to b-2. Production after improvement is precisely following actual shipment. Inventory reduction is easy to see as represented by the vertical bar; it shows a reduction from 302 to 169 at the end of the

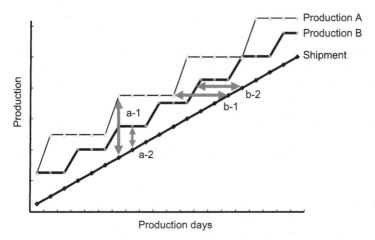

Figure 5.3 Accumulated chart

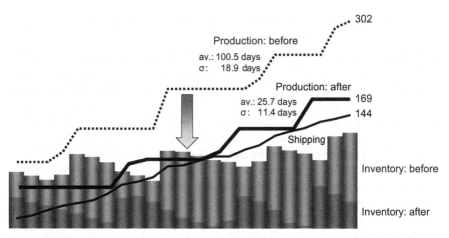

Figure 5.4 An accumulated chart practice

month. Then, production lead time decreased from an average of 100.5 days to 25.7 days. Another issue of U regarding productivity is that capacity depends on U.

A similar example is the cost reduction per products base and company level profitability. An important point for the relation is how to connect each product or parts cost reduction to the company's level of payment reduction. It's easy to understand but quite often, companies make this kind of mistake.

A company may use a higher level of manufacturing M and the P level may is also be good enough (100%), but the productivity results may still not contribute to the company's level of productivity without a higher level of U of manpower or machine U for the right level of inventory. Sometimes management is more inter-ested in the percentage of machine utilization because of the high monetary investment. You shouldn't have any difficulty understanding "economical manage-ment" with the following example. When you buy a luxury car like a Rolls Royce, just keeping the car in a garage will mean never driving it on the road. It is a cheap method because your invested money sank the cost and devalued your investment. Figures 5.2–5.4 further illustrate the power of these management techniques.

General improvement issues of U are:

- Tools and materials should be located to fixed areas and in suitable condition, including spare subjects.
- Fixed periodical maintenance of tools but machines should be set and kept in the best condition.
- Unavoidable turbulence should be accepted and prepared for with maintenance and machine utilization.
- Supplied components should be inspected preceding assembly.
- Practical scheduling is based on current production conditions.

5.3   Methods and Performance Meaning with Standard Time

Methods in engineering standard time include shop and/or workplace layout, number of workers (manning), machine or facility operating speed, sequences of operations, material handling, jigs and fixtures, operation procedures, and motions (See Figure 5.5). P includes lost time due to workers, management, and unavoid-able situations. Losses in production stages are unreasonable interruption of work, low motivation, low effort level, defect products due to carelessness of workers, disregard for standard operation procedures, low skill level, and so on, these are workers' responsibilities. It means the FM should supervise and instruct to im-prove/reduce workers' loss time. These kinds of losses are more than 80% of total loss time in shops.

Next is management's or support staff's responsibilities such as idle time due to materials, machines/facilities, and quality defects for out-of-control reasons. This is less than 20%. The last one is unavoidable nonworking time such as power failure, accident, disaster, and labor union activity. These make up a negligible size of nonworking hours. FMs' responsibility with regard to loss time should be controlled with a performance control system based on engineered time standard.

Look at Figure 5.6 in which workers employ a few different methods before setting standard time. A standard method as a basis of standard time is the mini-mum time-consuming operation among practices at the shop. Then the methods change to time valued as standard time. This is not an improvement; it is wel-comed as a change method. So, measure of method effectiveness is the measured difference of time value, and the shortest time value of method is normally the one best way. Then, P is measured by worker results to keep standard methods. Vari-ance of actual time value means difference from standard methods in time value. This is very important to improve productivity based on the engineering way of

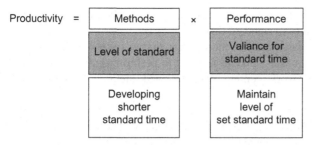

Figure 5.5 M and P meaning with standard time (Sakamoto 1983)

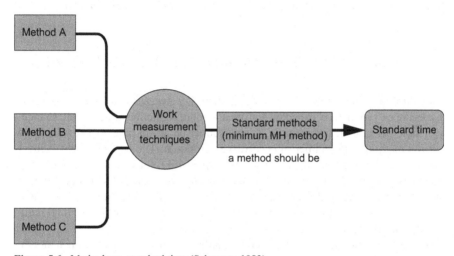

Figure 5.6 Methods vs. standard time (Sakamoto 1983)

thinking. Methods can evaluate those advantages or disadvantages with a time value of standard time, whether those methods can be done by the operator or not at shops at all. Evaluation of methods can be done in a cool or static way to rate the conditions or results at shops. Methods change for productivity improvement is an activity to develop shorter standard time work methods. So, even the better methods with standard time evaluation do not follow standard time at shops sometimes. This is why workers' P has to be measured with standard time.

P is measuring the difference between actual time and standard time in percentages. A 100% P rating means the worker followed standard time within the actual operation. Additionally, the worker followed standard methods as a basis of standard time including standard pace of working or others that encompass the standard time "definition". A rating of lower than 100% for P means poor actual condition compared to standard time contents; more than 100% means actual operator's method was better than standard time contents coupled with higher working pace. Productivity results can be measured with product of level/difference of more than one method and actual following condition to standard time as

P. Performance results can fluctuate depending on which method the worker employs and whether he adheres to standard time.

B. Niebel wrote: "Standards are the end of time study or work measurement. This technique establishes a time standard allowed to perform a given task, based on measurements of the work content of prescribed methods" (Niebel and Freivalds 2003).

There is a common way to set targets and evaluate improvement, simply measure and compare two methods of the current improvement. There is no objective fairness, as this is very much a subjective way. There is a long history of developing improved work measurement techniques while maintaining theoretical background for keeping reasonable accuracy and fairness (Bayha and Karger 1977).

5.4 Meaning of Standard Time on the Productivity Dimension

Let's introduce an example of U in which there is a necessity for changeover time that reduces production capacity and creates nonproduction time. Total changeover time per month, for example, is changeover or set-up time as a standard that follows methods and number of occurrence per month. The method of changeover can reduce standard time with improvement of SMED, for example. This issue of changeover as the U dimension is evaluated by its effect as M dimension. Another part of changeover, total frequency of changeover, is the outcome due to production planning and control. So, it is possible to measure for the effect of total number of changeover times.

There is no standard of a reasonable number of changeover times per month; however, the desirable number of time is as small as possible. This is why a practical measurement of this matter of U dimension is measured with the difference between actual results of total changeover time and total changeover time in a benchmark month. It is possible to measure quality and facility maintenance matters with a standard time change. One operator added for standard manning is allocated in order to prevent quality problems, for example. Then allocated manning difference is possible to measure as standard, so the difference of that standard time is effect of this matter of productivity.

Similar measures can be effective for preventive maintenance of machines and facilities.

Let's introduce a new and meaningful relationship of M and P for productivity improvement. It is also a significant point of view for effective management. I can explain this relationship with common tools in kitchen use. Water is in a bamboo basket or metal bowl while in the water. Water leaks when the bamboo basket lifts up from water but there is no leaking from the metal bowl. There is a simple example of M & P. The basket or bowl is the method. Water gets into them when they are in the water but water leaks out with the bamboo basket. In order to keep water in the bamboo basket, it has to be sealed to protect against leaking water; this protective action is P. Water cannot be kept in the bamboo basket (M) without protection (P). Like this, no method can guarantee its (M) effect without protection (P) issues. Methods effectiveness occurs with measurement and evaluation with standard time whether meeting those time values or not. This is the difference of methods effectiveness and it is measured by standard time. The difference of standard time value is the difference of methods effectiveness. How much to meet those methods are measured as P level. Difference of P level is not dependent on methods themselves. This is why actual methods contribution such as M × P cannot manage without measuring P.

U means taking water or not with a bowl or a protected bamboo basket. Doing an operation such as a method or not belongs in the U category for productivity. Before M and P, the necessity to do or not to do such a work has to be decided upon.

Figure 5.7 shows which working areas of a factory are effective to implement WIP. Each working area has its own cost of production, such as JPY 23,295 for a product at raw material stage and JPY 32,400 after press work. There is a common way of thinking today that minimizes the importance of production lot size. This is not a problem exactly, but there is a missing equal-value point of view. Six is set as the final assembly line, for example, and any working area is also set at six in lot size. There is no special reason to keep the same number of lot size for all working areas. At a metal workshop, 2–3 min are spent for punch press work for six pieces then set up/changeover every 15 or 20 min. Equal value for raw material is 12M JPY compared to a product 100M JPY, 22M JPY for after-press work. This means about ten times inventory for raw material is the equal value of

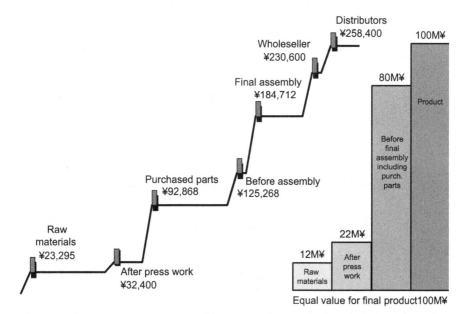

Figure 5.7 Cost from raw materials to finish products

five times before final assembly. Ten times for raw material, five times for before final assembly are acceptable sizes of inventory. The result is reducing lead time because there is enough WIP before final assembly, before press work and so on. Discovery of a solution of this kind belongs to the U dimension.

References

Bayha F, Karger DW (1977) Engineered work measurement. Industrial Press, New York

Helmrich K (2003) Productivity process: Methods and experiences of measuring and improving. International MTM Directorate, Stockholm, Sweden

Krick EV (1965) An introduction to engineering & engineering design. John Wiley & Sons, New York

Niebel B, Freivalds A (2003) Methods, standards, and design, 7th edn. McGraw Hill Professional Books, Boston, MA

Sakamoto S (1983) Practices of work measurement. Japan Management Association, Tokyo, Japan

Chapter 6
Methods Design Concept

Chapter 6 defines MDC as a unique approach to increasing productivity with the method functions. Any ideas or improvement results appear naturally; there is no magical way to find unique ideas. Ideas are already embedded in the brain of designers as untapped potential. Those ideas materialize with a specific attitude in designing steps. Those are included in the MDC steps. Any operation has functions; functions provide the purpose of the operation. Functional analysis includes the functions of BF or AF. A function has several methods to meet the function. It depends on OP definition of the work area. Any unique design starts with clear specifications of design at the beginning. MDC calculates Kaizenshiro as the theoretical target. It also considers "real gain" as designing new methods. Design areas in MDC are manufacturing methods, manufacturing systems, and management systems.

6.1 Application Results

6.1.1 Improvement of Workers Number

MDC results are shown in Table 6.1 representing an almost 50% manning reduction or more than 150% productivity improvement without changing the production cycle time. It is easy to reduce manning with an increase of production cycle time. Only one worker possibly assembles cars, for example, if the production cycle time is acceptable at a longer rate. There are 250 workers on a typical car assembly line with 2 min of cycle time. The number of workers can be reduced by 50%, for example, if the line cycle time is increased two times the current rate, such as 4 min. Meanwhile, there is a possibility of a 50% reduction of manning. These results of productivity improvement of 200% (two times) should be realized as unique results of manning, not productivity improvement, if mechanization and/or automation can be installed well, but the acceptable capital investment and development of mechanical engineering may be too high.

A few examples of a module for MDC object working areas are shown in Table 6.2. There is not only productivity improvement percentage but also expense for change, increasing BF, and decreasing AF percentages.

Table 6.1 MDC results in a few companies (Sakamoto 1992)

Company		Productivity improvement (%)	Number of workers	
			Before	After
A	Cars	205	1,314	642
B	Telecommunication devices	132	699	529
C	Machineries	177	44,581	25,202
D	Foods materials	177	266	150
E	Foods	176	590	335
F	Sheet glasses	169	1,484	876
G	Car parts	178	1,683	948
H	Aluminum process	166	445	268
I	Cameras	192	230	120
J	Foods	164	245	149

Table 6.2 MDC results in small modules

Modules	Improved results			After MDC – work amount of model				Before MDC – work amount of model			
	Reduced manning	Increasing productivity	Expenses for change			BF	AF			BF	AF
	no.	%	(1,000 ¥)	no.		%		no.		%	
A welding of stocker	2	213	750	3	2.343	76	24	5	3.291	57	43
B assembly flash tank	4	201	708	4	3.988	60	40	8	6.78	39	61
C store box assembly	3	182	1,240	3	3.297	66	34	6	4.974	55	45
D refrigirator sub. assembly	3	192	1,300	3	3.129	65	35	6	4.88	50	50
E inner-outer assembly	4	192	1,700	4	4.165	63	37	8	5.634	49	51
F washdisher sub. assembly	4	161	1,610	5	5.58	68	32	9	8.169	47	53
F under ref. assembly	2.5	202	400	2	2.229	62	38	4.5	3.722	50	50

6.1.2 Improvement of Set-up Operations

Consider an example of reduced changeover time from 5 to 0.5 min at a steel facility in Sweden. This production line produces a small lot size of products so it was important to reduce the number of changeover opportunities by production planning and time efficiency for changeover. MDC is applied to the latter subject of reducing changeover time. Figures 6.1–6.3 are before MDC application and after. The special tool developed through MDC is in the category of innovation. The original method was walking to the machine, unscrewing a heavy tube, transporting it to a storage place, transporting a new tube to the machine, adjusting the machine, and mounting the new tube. The work content is 45% BF, 55% AF. The new design method is to unfasten the tube, rotate it to a new size position, and fasten it with an attached small lever.

There is no heavy tool that is transported by crane; instead, all different sizes of tools are arranged in a circle and fixed on the machine. Set-up operation is to just rotate it by a maximum of 180°, and it requires only 0.5 min. This production line completed single-minute set-up many years ago but management's interest is to reduce set-up time even more. Their requirement meets the MDC that starts from a blank sheet for finding new methods.

Figure 6.1 An example: MDC for set up at SKF steel (result)

Figure 6.2 An example: MDC for set up at SKF steel (before)

Figure 6.3 An example: MDC for set up at SKF steel (after)

6.1.3 Sequence Analysis for Mechanized Machine

Sequence analysis is primarily used for mechanical processes. The purpose is to describe each step for a piece of mechanical equipment, including slack, overlaps, and intermediate movements. The analysis provides a clear picture of the relationship between the movements in the different parts of the process. It is not uncommon for there to be significant slack between different movements. By reprogramming the equipment it is often possible to increase capacity considerably. To sum up, the analysis provides an excellent visual image of how the movements are linked; which are dependent and which are independent.

Figure 6.4 is an example of MDC application for a mechanized bottling machine in order to increase machine capacity. The cycle time before MDC application was (0.06 min/bottle = 17 bottles per min), its BF and AF were 47 and 53%, respectively, of cycle time. The designed cycle time was 0.04 min/bottle = 25 bottles per min; increasing capacity by 150%. Key parts of increasing the capacity were achieved by reducing transfer time of the machine and waiting time for different mechanisms of the machine, and a few movements were changed to move in sync. As you can imagine, such change of the machine does not require higher cost. The machine used to comprise all movements as one series in industry. Today, the MDC result comes from finding the possibility in all movement. This concurrent movement is often missing when mechanical engineers design machines. Industrial engineers, in contrast, take concurrent movement into consideration. This change of movement sequence only requires rearrangement of a control box and requires a very small expenditure. Figure 6.5 is another example: 2.5 s improving cycle time from present cycle time of 6.39 s.

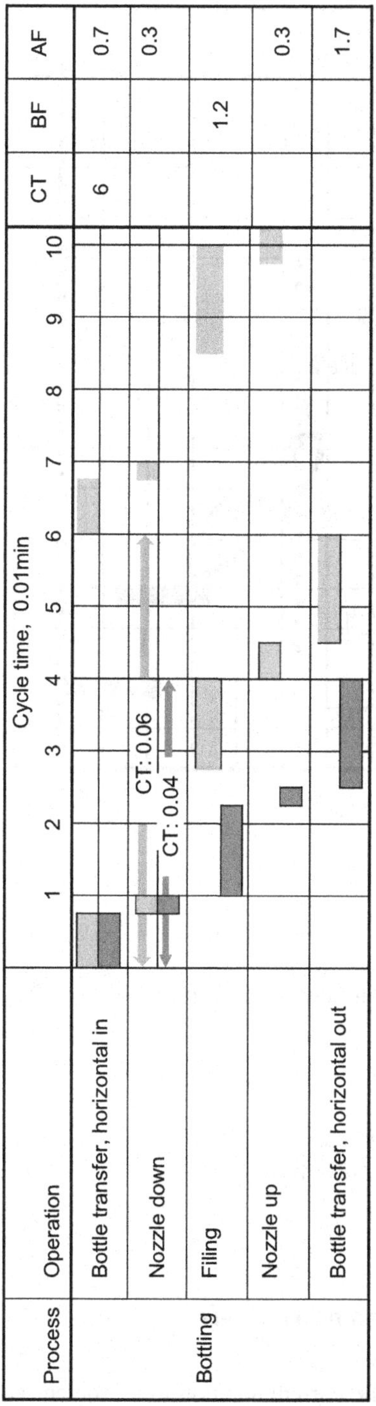

Figure 6.4 Sequential analysis: cycle time reduction of bottling operation: company C

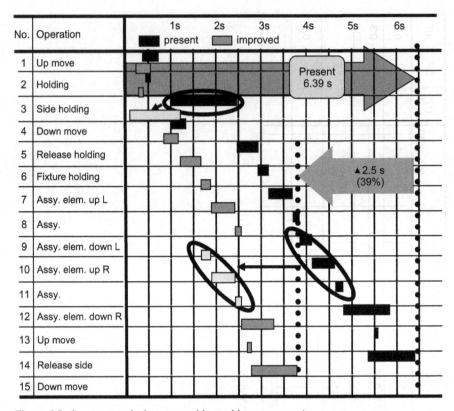

Figure 6.5 Sequence analysis: an assembly machine: company A

6.2 Fundamental Points of MDC

There are a few points to follow when designing new methods for unique results. Those are:

- Disregard or leave behind current methods. Think from a blank sheet.
- Functional analyses with OP definition.
- New ideas through creative or innovative thinking.
- Faithful reflection of management requirements or policy.
- Target-/Kaizenshiro-oriented design.

6.2.1 Disregard or Leave Behind Current Methods

Thinking from a blank sheet reinforces that "the current solution to a problem is not the problem itself" (Krick 1965).

6.2.2   Functional Analyses with Output Definition

Any operation has functions; functions provide the purpose of the operation. Functional analysis includes the functions of BF and AF. It depends on an OP definition of the work area. An operation can demand change in classification of BF or AF, according to the OP definition.

6.2.3   New Ideas Through Creative or Innovative Thinking

Waste or small changes of current operations cannot be an objective of MDC design. The four principles of methods improvement are powerful directions to find such ideas.

6.2.4   Faithful Reflection of Management Requirements or Policy

Design to meet management requirements has to be the basic point of MDC. Method changes are limitless in any shop, but real results require expenditure for implementation and an acceptable time schedule. Everything has to meet management requirements. Note, however, that management cannot indicate their requirements; industrial engineers must investigate using a feasibility study and show the results.

6.2.5   Target/Kaizenshiro Oriented Design

Any unique design starts with clear specifications of design at the beginning. MDC calculates Kaizenshiro as the theoretical target. When designing work methods, do not forget the target.

6.3   Features of MDC

6.3.1   What Is the Objective of Applying MDC?

Shift hours are the contents of working and idle or nonworking hours. Three levels of improvement mentioned earlier connect with MDC. Work simplification reduces idle, nonworking hours, or waste. Improvement includes some AFs. Innovation within MDC is the opposite approach; it has no interest in reducing idle time, waste, or part of AF work. MDC increases the share of BF in work contents. Nonworking hours or waste are not studied; methods design concentrates on finding solutions that can increase the share of BF in working hours. It is known as the outside-inside, instead of inside-outside, approach.

Note the feature of MDC, which helps to get such a unique productivity improvement: a limited objective-oriented approach. What is your objective in applying MDC? Improving productivity by reducing manning, shortening cycle time, and increasing production capacity. It is very important to recognize an objective and to concentrate on this single objective. Corporate performance is also an important basis for doing productivity improvement. Let's consider an example that illustrates how easy it can be to employ the practice of productivity improvement. Entering production floors, finding waste and then reducing it; calculating the improvement effects such as current cycle time minus reduced time with improvement, and multiplying the number of production volume per month or year. Then you can get total reduction time per month or year. Does this result in a real reduction of labor cost or improve productivity? This is not to say that there is not any effect, but it is not this calculation that matters. The first observation is that waste which is included in labor cost. The second is the possible reduction of actual labor cost based on the calculation results.

But again, how much actual effect on corporation performance will these steps cause? A 1-min reduction of cycle time and 30-min reduction per day is calculated. The objective is now to utilize the 30 min of improvement to lead to a reduction in shift hours. The answer does not easily reflect those improvements of time results into shift hours.

There are a few steps in designing new methods, such as reducing cycle time, reducing allocation of workers, and quality issues. A wrong way of thinking is that the design objective is not limited. Again, a single objective is recommended. Otherwise, unfruitful discussion may happen: "Cycle time can be reduced, but there is a question regarding quality". The recommendation is to just design a new method to meet cycle time, for example, and think about quality issues later.

6.3.2 Designing New Methods with an Engineering Approach

MDC is the design approach to find new methods instead of simply reducing waste. Design is based on the engineering view and defined to change IP conditions such as necessary work contents: time; OP results, such as number of manning; production cycle time, and under set limitation such as allowable expenses/cost; necessary weeks or months and any requirement from marketing, product design, for implementing new designed methods. Designing process and contents are similar to products design.

When you use traditional methods of improvement such as work simplification, you will be interested in the distance for getting and placing parts, utilizing both hands, and improvements without changing main parts of jigs and fixtures. If you follow the MDC, however, there is more opportunity to find a totally different way of improving methods that is not just following the current basic operation meth-

Figure 6.6 Work simplification
vs. MDC

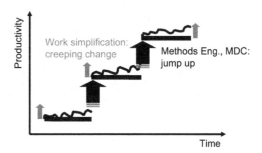

ods. The point of this difference is whether you concentrate on the BF or not. This is why the MDC will lead to a jump to a new, higher platform as illustrated by Figure 6.6.

The engineering approach sets targets at the beginning and searches for ideas to meet the targets. Continuous improvement does not deny results, but it implies an endless approach because reasonable targets are not set before finding improvement. This is why this endless improvement has weaknesses; it cannot evaluate a high level of improvement to meet management's or the company's requirements. Generally, any engineering approach of problem solving is to set specifications first. Machines do this (otherwise, they are not valuable). Price is an example; product design engineers' creativity is limited by the market price.

An improvement target must be set as high as possible and by the improvement project itself rather than through the copying of other projects' attainments. The MDC may not be applied to current methods since innovative improvements do not work through analysis of current methods. In place of this approach, suitable methods for defined OP should be created so that innovation results. Having gone through MDC, innovation will reach the ideal level.

To leave behind current methods or simple definitions of IP and OP that are based on current conditions will create a fresh slate, that blank sheet of paper.

6.3.3 Focusing Function of Work Contents

One operation is no more than one of a number of alternative operations to achieve a given function; accordingly, rather than focus on the work or operation itself, importance is placed on studying *functions* as the desired outcome of operations. On the contrary, a function has plural alternatives of methods to meet the function (Figure 6.7).

Within manufacturing, the operations and processes are means of converting from an IP status to an OP status. At a lower level are operational elements and motions, some of which contribute directly to meeting OP demands and others which support upstream work contents. These two categories have been defined as BF and AF. As you see in the MDC steps, defining the function of an operation

Figure 6.7 Function and methods

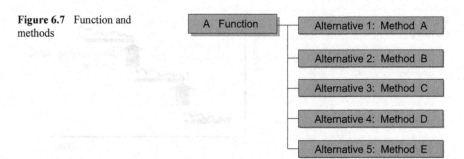

after setting work contents is a unique step. For example, if you fasten two parts with bolts and nuts at the beginning of the MDC steps, the next step is to define a function of the fastening operation. The function of the operation is "fasten"; then you can find another step is to create ideas to meet this function of fasten. There is the possibility of changing product design; attention to customers' requirements are also matters that do not change in methods design. These matters take time to change and approve. This is why I recommend avoiding those issues for methods design. A function has a few working methods to do. What jigs, fixtures, tools, sequence within other operations, or number of parts for handling have the possibility of meeting a function? The number of bolts, other methods of welding, and a hinge for snapping are alternative methods of product design but they require the engineering area's change that is suitable for methods design concentration. This kind of change will make fundamental changes, but they are not easy to get approval for from the design department. According to the practices of MDC, to challenge and find innovative ideas that concentrate on functions is better than just considering the methods themselves. Searching for ideas with a function-oriented premise guarantees a wider view. Free thinking is possible.

Thus, MDC increases BF work, resulting in higher productivity improvement and innovative ideas. Let's think about farmers' harvesting by picking fruit. What does "picking" mean? In this case, there are several methods: vibration (shaking, sound), pulling (human hands, suction, mechanical), cutting, burning (electric resistance, flame, solder gun). A purpose such as function of this harvesting is obviously not meant for examples like vibrating, pilling, cutting, burning, but meant for separation (Krick 1965). "Separation" is a single function and there are different operations for the function of separating fruit from trees. It is important to find a creative expression of functions/purposes when observing current methods. The expression of functions is not tied up in the current methods.

The innovation approach is to study the BFs of the operation, to develop a completely new method of performing them without regard to the current method, and to accomplish the BF in a minimum amount of time. This is the approach that would lead to the development of a nailing machine to replace a conventional hammer and nail operation.

With previously described types of methods improvement, you may have a large number of very small ideas that can only take you so far. The approach must be changed, or efficient and creative ideas will not be found to raise the potential for efficiency. This level is achieved through the design approach, such as the MDC.

One is eager to find different methods without paying much attention to the current working methods. There may be simple improvement ideas if you just observe current methods for picking fruit, for example: cutting fruit quickly with scissors in which there is the possibility of using both hands for two cutting tools simultaneously, or by shortening the distance between the tree and the collection basket, if you do not search ideas for the function of separation.

What methods should be used to separate the fruit from trees? Your ideas will expand better if you do not limit yourself. Krick says that "current methods themselves are not problems". This is a very important point of view to find new creative ideas. There is no need to consider if ideas are "practical" or not when searching for ideas.

As you can see in the MDC table, there are specific ways to describe methods for a function. We normally think we do many types of operations within the organization but those operations are very much limited in functions. Again, think about the farming example. Searching for scissors that quickly cut fruit or using both hands simultaneously, reducing distance from tree to harvest baskets. What methods are there for a function to separates fruit from trees? When you use traditional methods improvement such as work simplification, you will be interested in the distance for getting and placing parts, utilizing both hands, and improvements without changing the main parts of jigs and fixtures. If you follow the MDC, however, there is more opportunity for discovery.

Words for describing functions are limited such as fasten, change places, change positions, fixing position, or connect as one piece or element. There are specific ways to describe methods for a function. We normally think we do many types of operations within the organization but those operations are very much limited in functions. It is recommended to prepare a list of common descriptions as a quick reference. Table 6.3 shows the difference.

Rather than spending time trying to figure out how to make a particular operation more efficient, it would probably be intrinsically more effective to consider what the objective of the operation is. In the automobile industry, for example, the basic operations for vehicle assembly will be the same in any factory. However,

Table 6.3 Example of methods and functions

Function description	Methods description
Part A fix to basement	– fix with screw
	– fix screw through hole of part A
	– insert part A to body with friction of plastic
	– put part A and fix with other part B put on A
Change place from X to Y	– assembled part put on handling carrier one by one
	– completed subassembly on the conveyor

the time spent in producing a particular product and the breakdown of time spent on a single shift can differ according to the factory, company, drawings, production system, the supervisors' instruction, or the performance of the workers themselves. The contents of this fundamental operation can be considered as resembling the BF of the MDC.

If we consider the problem of joining two items together, here are a few options: glue, nut and bolt, welding, and hinging. Rather than taking an approach based on preconceived notions about the process, greater improvements can be realized by questioning the process and asking why it must be nut and bolt. This is to say that an approach in which the objective is neglected while attention is paid only to function should be avoided. Improvements will be made only on current processes where this is the case.

6.3.4   New Methods Are Easy to Implement

There are two reasons for implementing easy methods. One is that MDC is an objective- and target-oriented approach and its ideas have no limits, like a never-ending story. MDC sets a target at the beginning step. Searching and creating ideas continue until the target is met. It finishes when the desired improvement effect is identified.

Another reason is low-cost improvement through searching and creating ideas based on the company's individuality. Higher cost and/or capital for changing methods might require a large scale of change from current methods. Cheaper methods do not require a longer term to implement.

6.3.5   Design Company Owns Original Methods

Improving productivity without capital or a large amount of expenses is a tactic of MDC. The objective methods or machines/devices are not available in the market. Just hiring them from the market shows a poor level of searching out ideas. You must create new ideas for your own company. There is a secret to acquiring these ideas. One useful tool is to study the four principles of improvement: eliminate (E), combine (C), rearrange (R), and simplify (S) and ask yourself which of them is applicable to your purpose. And those machines and devices you're thinking about getting? Anybody can purchase them! This means that any advancement from that machine or device has no guarantee compared to other competitors. Even with effective methods from outside resources, the effect for profitability should be considered. Reduced labor cost through high investment means reducing profitability. This is why MDC means developing new methods without higher investment.

6.4 Areas of Design

MDC covers three areas:

- Manufacturing methods;
- Manufacturing systems; and
- Management systems.

6.4.1 Manufacturing Methods

First, there are manufacturing methods that involve machines, facilities, and tools. There are many types of machines, for example, and the problem for engineers is which machine to use for each particular function within the methods. There will be several machine tool combinations for any particular function of production methods.

A company in which MDC was adopted developed a totally new shape of a cutting chip for a machine tool. The design target of the cutting operation was to reduce the operation cycle time from 31 to 18 s, a 42% improvement. The industrial engineer tried to find an effective cutting tool that was both fast in removing metal and could be changed quickly. Because adequate cutting tools were not easily found on the market, the engineer asked a world-famous cutting chips producer if they could produce one. The producer replied that it would be very difficult, but they did succeed in developing such a chip. This is a simple example where the MDC process did not just accept improvements attainable through easily available existing tooling. Instead, the original design targets were met by trying and succeeding in the development of their new tool. The tool also had another feature which was easy to set up because of the special shape of the design of the chip. Before improvement it took 10 min. Now it takes less than 2 min.

6.4.2 Manufacturing Systems

The second area covered by MDC is manufacturing systems. This involves whether to use batch production or synchronized production; whether to choose individual, group, or line working; and the design of layout.

Questions involved in the manufacturing system are: Why do you adopt a continuous flow line for particular products? Why do you install a certain number of robots in the production lines? What is the relationship between workers and machines? Do you apply a straight or U-shaped line or layout? There is a lot of room to create new and effective systems. Manufacturing people prefer to install continuous, long, straight lines in their plant. But have they researched why their solution is more effective than another solution? The answer is usually no, even though everybody agrees that it is necessary. The question is how to do it and not what to do. One of the practical solutions is to change the approach from the experience-based approach to a design approach as in the MDC.

6.4.3 Management Systems

Finally, the third area, management systems, involves developing systems that can make effective use of workforce motivation and performance management.

Even if good manufacturing methods and systems are adopted, there is still one other point of design, the management system. How do you manage if unusual things happen? We have to be prepared for any uncertainties on the shop floor before the total system can work well.

A good example is industrial engineers preparing a production line. They calculate the loss in balancing a line compared with the cycle time of each workstation. This kind of line balancing is called static line balancing (SLB). The actual practice of line balancing is quite different, however, depending on the workers' performance, interference, and work mix. Hence, industrial engineers must use more complete techniques such as DLB. But how do you manage unskilled workers who are assigned to a particular workstation, unusual material conditions, and machine conditions? The MDC has been designed to leave a choice of alternatives of manufacturing tools, manufacturing systems, and management systems. SLB is a line balancing calculation based on standards with no performance fluctuations and differences. In contrast, DLB is an actual line balancing in shops. Logically, differences between SLB and DLB always happen because performance fluctuates depending on workers and even cycles per worker. Therefore, it is important for managers and supervisors to take actions to resolve the balance loss caused by the difference between the actual DLB and the higher SLB. Such WIP between workstations suggests the possibility of further manning reduction. Managers and supervisors must find this possibility and take actions for improvement.

Two types of knowledge concepts by scientist and philosopher Michael Polanyi are tacit knowledge and explicit knowledge. This concept can be explained with a DLB and SLB background. Explicit knowledge is knowledge that has been or can be articulated, codified, and stored in certain media. It can be readily transmitted to others. The information contained in encyclopedias (including wikipedia) is a good example of explicit knowledge. Explicit knowledge is codified, and can be precisely and formally articulated, is easy to codify, document, transfer, share, and communicate. Its ready accessibility has lead to many ways of using it as a management tool. Explicit knowledge is increasingly being emphasized in both practice and literature, as a management tool to be exploited for the manipulation of organizational knowledge.

It is important to understand that he wrote about a process (hence tacit knowledge) and not a form of knowledge. However, his phrase has been taken up to name a form of knowledge that is *apparently wholly or partly inexplicable*. With tacit knowledge, people are not often aware of the knowledge they possess or how it can be valuable to others. Tacit knowledge is considered more valuable because it provides context for people, places, ideas, and experiences. Effective transfer of tacit knowledge generally requires extensive personal contact and trust. Tacit knowledge is not easily shared. One of Polanyi's famous aphorisms is: "We know more than we can tell." Tacit knowledge consists often of habits and culture that we do not recognize in ourselves. In the field of knowledge management the concept of tacit knowledge refers to a knowledge which is only known by an individual and which is difficult to communicate to the rest of an organization. Knowledge that is easy to communicate is called explicit knowledge. The process of transforming tacit knowledge into explicit knowledge is known as codification or articulation.

Tacit knowledge is generally described as: subconsciously understood or applied, difficult to articulate, developed from direct action and experience, shared through conversation, story-telling, *etc.* Since it is personal and context-specific, it is difficult to articulate.

If it is easy to find alternatives, all alternatives should be compared as thoroughly as possible. If it is difficult to find innovative ideas with the usual easy approach, then use the MDC application; mechanical engineers, electromechanical engineers, production engineers, and industrial engineers are almost always experienced enough to develop effective manufacturing systems and manufacturing methods. Their experience concerning management systems for the workers, materials, and machines, however, are poor because there is no simple step-by-step approach to management systems.

The importance of performance management is particularly noteworthy. Standard time and standard operational procedures are useful for effectively implementing newly designed methods, finding and retaining workers, supervising workers, and getting workers to understand the newly designed working methods, by not only standardizing procedures but also by indicating standard times based on MTM.

6.5 Development Steps of MDC

MDC steps are summarized in Figure 6.8.

Step 1 Setting Modules/Structuring

Step 1.1 Setting Modules
Choosing the right improvement area for MDC is very important. There are two points in considering which modules to concentrate on: first is the number of workers in the module and second, the job family. The number of workers in a module is important because the module must be easy to identify and define. It is difficult to visualize and understand the structure of a module if it is larger than

15 workers. The industrial engineer has to visualize the borders and image of the modules as the operation methods are designed. If in doubt, however, it is better to have large modules rather than too small. Too small a group will limit the possible improvement potential. A group of three workers gives a maximum potential of only three workers, while a bigger group creates a wider range of possible improvements, and hence the potential is greater.

Rather than retaining existing work blocks (modules), totally new modules must be established through new design. Ideally, modules should be established by considering the operation and its methods. Once a single MDC is developed, its results can be applied to multiple modules. By skillfully defining modules, points that appear to be different under the current organization or line structure can take on partial commonality or similarity (not necessarily identical, but similar). It is therefore desirable that modules that are as similar as possible be established.

Let's look at an example of module setting: suppose there is a machine shop and beyond this shop there is another shop, the finishing shop of the machined processed part. They are on separate shop floors, but these two shops are in a very close functional relationship. This means the product flows from the first shop to the second. In this case, both of the shops should be seen as one module. If they are seen as separate modules, the IP definition of the second module is just process-machined materials and the OP is parts that are ready to use in assembly work. But if they are seen as one module, the IP is raw materials and OP is parts ready for the assembly shop. The necessity for finishing is not important for designing new working methods.

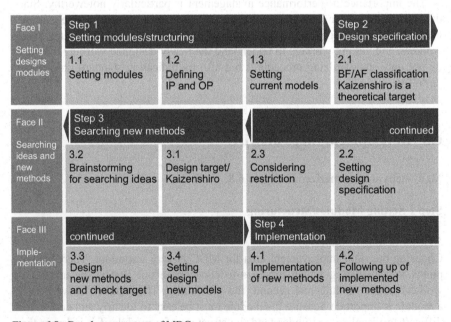

Figure 6.8 Development steps of MDC

A similar example is circuit board auto mounting, inspection, and adjustment. Those are separate shops but should be considered as a single module for the MDC. Inspection and adjustment work themselves are not the purpose of the work. The ideal definitions of IP and OP are various kinds of supplied parts and mounted printed circuit boards ready for assembly.

There are three points when you set modules for successful developing of MDC steps according to MDC practices: the objective of MDC development, the definition of IP/OP can change, and BF.

What would you like to apply MDC to? Typical objectives are reducing manning, cycle time reduction of machines, reducing set-up/changeover time, and so on. Especially for the objective of reducing manning, expected results of reducing the number of operators vary by modules setting as shown below. That means the definitions of IP and OP vary depend on module setting. This means the definition and classification of BF is automatically changed.

Step 1.2 Define Input and Output
A module is the transfer from state (condition) A to state (condition) B. "In any problem there is an originating state of affairs the problem solver seeks a mean of achieving; call it state B (Figure 6.9). A solution is a means of achieving the desired transformation. A problem to which there is only one possible solution is rare indeed; for most problems there are many alternative solutions, many more than there is time to investigate. In addition, a problem involves more than finding a solution; it requires finding a preferred means of achieving the desired transformation."

"It is difficult to imagine a problem in which there are no restrictions on solutions. A restriction is something that must be true of a solution...as an engineer you should be skilled at identifying the basic characteristics of the problems you are to solve" (Krick 1965).

Figure 6.9 Methods: change from A to B

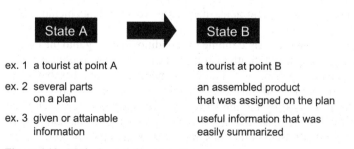

Figure 6.10 Methods: state from A to B

Figure 6.11 Change IP to OP: black
box

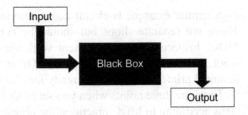

The MDC steps for developing new methods are indicated in Figure 6.8. Setting modules objectives according to methods design objectives is the first step. Methods is changing the process from condition A to condition B. The example of condition image is shown in Figure 6.10.

The method is a black box, meaning any method to make a change from condition A to condition B is acceptable, if not perfect. Condition or state A is the explanation before a black box as certain methods, then condition B or state B is the explanation after method (Figure 6.11). With regard to condition, IP is a condition before the processing method and OP is after processing. It is not simple work to set a reasonable module when MDC is applied for the current working areas because the next step of IP and OP definition could be different. An example is MDC for reducing the manning number at parts mounting of printed circuit boards. There are two main workshops for mounting printed circuit boards, such as mounting parts on board and 100% of inspection. Two ways of module setting are possible. The one is mounting parts on board (module a) and after this inspection (module b).

In this case, the first module OP is defined as a board with just-mounted parts and then that is why the IP of the next module is a mounted board and then after an acceptable quality of boards through inspections. A single background of this module setting is based on the current organization, such as a mounting shop and inspection shop. There is a manning number of many workers at present. If module setting is done like this in two modules, current inspection work is automatically accepted without any questions of necessary inspection. It is a kind of fault of developing new methods. A kind of restriction is accepted automatically, that is, two shops are accepted as one. Another module setting is inspection free of mounted boards. IP is boards and parts; OP is mounted boards for final assembly. Workshops based on the current organization have no meaning in a module setting.

Methods work as a function to change IP to OP conditions; methods convert the process regarding IP and OP conditions. This is why current work methods themselves have no significant meaning; significant issues are the definition of IP and OP. Depending on those definitions, necessary work contents/methods are changed automatically.

Defining IP and OP means clearly defining the conditions coming into the model and those going out. During this definition process, the tendency is to confuse the IP or OP of products and components with that of the methods. Remember that what we are designing here is the work methods. Therefore, the definition covers what conditions (IP) are processed by the model to produce what conditions (OP). In order to avoid confusion, we make it a rule to define products and components, as well as methods IP and OP. Both definitions may be written in the operation sheet shown Figure 6.12.

FORMAT B

MDC SPECIFICATION

	PRESENT MODEL	KAIZENNSHIRO	FUNDAMENTAL DESIGN	DETAILED DESIGN
BF	%		%	%
AF	%	%	%	%

INPUT

1. inspected package
2. all inspected parts those are indicated on the drawing

PRODUCT

OUTPUT

1. assembled unit of required quality

1. all parts are supplied with a kit by overhead conveyor
2. jigs and fixtures are prepared

METHODS

1. assembled and inspected units are sent to a packaging shop

PRESENT: references
production volume: 540 pcs/day
cycle time: 28 seconds
manufacturing system:
continuous flow line
number of workers: 5 workers

NEW DESIGN: specification details
production volume: 540 pcs/day
cycle time: 28 seconds
manufacturing system:
any systems are OK
number of workers: 5 workers
follow to the KAIZENSHIRO

RESTRICTIONS

present

new methods

Figure 6.12 MDC format B

A module or working method is a process to change from IP to OP. Methods are changing processes between a given condition as IP and demand results as OP. There are different definitions of IP, and OP current working methods. The simplest and worst example of IP and OP is just writing up details of current methods. But the important matter at the moment is to improve current methods themselves. The design target is to develop absolutely new methods for current modules. IP is before process of the modules and OP is the results. This definition differs with restrictions, for example. Those definitions are to follow current conditions if you cannot leave current conditions.

Assume the process to be developed is to go from city A to city B. When the process is expressed as "to go from A to B" it leaves scope for creativity. If, on the other hand, the process is called "how to reduce the time to go by train from A to B", the phrasing imposes severe restrictions and limits creativity. A general definition of IP and OP is as follows:

- IP: Power or energy put into a machine or system for storage or for conversion in kind, or conversion of characteristics with the intent of sizable recovery in the form of OP.
- OP: Power or energy delivered by a machine or system for storage, conversion in kind, or conversion of characteristics.

Step 1.3 Setting Current Models
Defining models means defining work contents, and confirmation of the current model. Levels of work content for defining the current model are shown in Table 6.4. It depends on the size of the modules or cycle time of modules, but normally levels two and three are recommended. The relationships of each measurement level are displayed in Figure 6.13. Such activity can be divided into a few processes; a process can be divided into a few operations, and so forth.

As shown in Figures 6.13 and 6.14, the BF ratio (%) depends on how much analysis levels are changed. This means the BF ratio of an operation increases

Table 6.4 Classification of work measurement

Level	Explanation
1	Motion: components of element, possible minimum measuring unit
2	Element: a few gathered of motion, measuring unit of DTS
3	Operation: operation is the only work unit which is defined clearly, and is the smallest work unit which can not be assigned/distributed for more than two workers
4	Process: a few series of operations at a work station; thoses are a series as contens of activity
5	Activity: a few series of processes for working process in order to complete function
6	Function: a part for assembled product or subassembled part, e.g., function includes all necessary activities

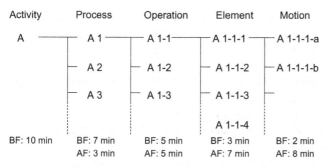

Activity	Process	Operation	Element	Motion
A	A 1	A 1-1	A 1-1-1	A 1-1-1-a
	A 2	A 1-2	A 1-1-2	A 1-1-1-b
	A 3	A 1-3	A 1-1-3	
			A 1-1-4	
BF: 10 min	BF: 7 min	BF: 5 min	BF: 3 min	BF: 2 min
	AF: 3 min	AF: 5 min	AF: 7 min	AF: 8 min

Figure 6.13 Relation of work measurement units

when the analysis level becomes small. As a result, Kaizenshiro increases when the analysis level is smaller.

It is possible to define an important point of view for finding reasonable work contents for use in designing new methods when you think about "pursuing the way it should be". This is not simple acceptance of current worksite status or improvement results with tooling and machinery that you desire, but the way it should be is totally different from both of them.

What is a model and what does it mean? A model is a simpler representation of the real world, described Moshe F. Rubinstein (Rubinstein 1975). The concept of a model is so fundamental to problem solving that it is present at all stages, from problem definition to solution. It is a concept characterized by uniqueness. The words and symbols we use, and the responses recorded by our senses, are all models. A model is an abstract description of the real world. It is a simple representation of more complex forms, processes, and functions of physical phenomena or ideas.

For an existing object or product being treated by method design, it is necessary to generate totally new creative models within the existing design because the operation currently exists. The purpose of confirmation of the current model is as follows:

- to understand the details of current operating methods;
- to understand the amount of time and work required by current operating methods; and
- to establish a benchmark in order to evaluate the effects of the newly designed methods.

A model should (although without details) be representative of the final shape and give the same impression. Analyze the current process. The process to be analyzed and improved is described in operation format A (Figure 6.14). (Note that the current process is to be described without any consideration of possible improvement potential already known at this stage.) Operation format A has the following principal columns:

Process: Give the designation of the whole process for which the module is made. If the process is large, it is advisable to subdivide it into operation steps and thereby describe the elements in the process. In this breakdown of work, no consideration is given to who performs the work. If there is any overlap between different elements, frequencies of elements, or the like, this is not noted. Standard times must be collected for the elements which are shown in the work unit (WU) column. The frequency of each operation is written in the work count (WC) column. Choose a suitable time unit for standards in order to make it easy to talk about the process in further discussion. The current model rather than current condition is required in operation format A. The model is a standardization of current operation methods, or a confirmation of operation methods as a desired practice under current conditions. The details include the content of the operation, the amount of time per occurrence (WU) (see Appendix 6, 25% Selection), and the required frequency per cycle (WC). The steps of the MDC unfold along identical lines. Consequently, the existing model should not contain the many day-to-day variations that occur. Exceptions to the general rules and specific conditions need not be included in the model. Instead, the model should be a simple representation of the process, not an analysis of detailed conditions.

Number of workers is calculated. On operation sheet A, the labor forces/work loads value are calculated as WU × WC, and it is divided by target cycle time (TCT). This conversion is performed for the labor force reduction purpose of the MDC application. It means any number of workload value is transferred in measurement unit from time value to number of workers.

For example, if I assume 0.50 min per cycle of operation (WU) with 10 times occurrences (WC) in a cycle and ICT is 0.25 min, then the numbers of workers on the TCT is $(0.50 × 10)/0.25 = 2.0$ workers for the work load.

It is a simple calculation but the method designers always think of any operation as the number of workers' work load when they would like to reduce the number of workers with the MDC. In operation sheet A, the labor forces/work loads value are calculated as WU × WC and it is divided by ideal cycle time (ICT). The ICT as used here stands for the production speed necessary to maintain the required work or OP. If the ICT is ignored, it is possible that the production volume will drop along with any reduction in labor power. General calculation or setting ICT is:

$$\text{ICT} = \text{total number of production} \div \text{net production hours}$$

Net production means prospected U, quality loss, and others if there are items for spending in the nonproduction category. It is a simple representation of more complex forms, processes, and functions of physical phenomena or ideas. Define current work contents regarding current methods or new settings of work contents regarding designing for real new modules. If you develop really new methods without current methods, this step is estimation based on industrial engineers' knowledge. Even applied to the current methods, this step is not simple methods analysis of current working methods. You have to seek the best possible methods as design models. You should find and define three subjects such as best working methods that are possible to do without improvement.

written by:	MODEL METHODS			module name				TCT: 60			/
	PROCESS			OPERATION			FUNCTION CLASSIFY		WORK COUNT	WC	WUX WC	NUMBER OF WORKERS
NO.	NAME	FUNCTION	NO.	NAME	FUNCTION		BE	AE	sec			
1	attach hock button	connect	1	picking panel base	change place			X	1.498	1	1.50	0.02
			2	disassy panel base and put aside	separate			X	2.026	1	2.03	0.03
			3	attach hock button	connect		X		1.440	1	1.44	0.02
2	attach mike assy.	connect	1	attach mike assy.	connect		X		5.472	1	5.47	0.09
3	attach packaging	connect	1	connect packaging	connect		X		3.744	1	3.74	0.06
			2	fasten screw	fixing			X	4.107	1	4.11	0.07
4	wire connect to speaker	connect	1	join connector	connect		X		2.160	1	2.16	0.04
			2	tidy wire	protect wire			X	1.962	1	1.96	0.03
					trouble							

Figure 6.14 MDC format A

MDC requires defining "should be", which is neither real current methods nor possible improvement methods. That means standardized methods, the method that is possible to do right now without any physical change, such as machines, devices, tools, and so on. This is why "should be" is the one best way under current production conditions. Actual contents of models is defining and standardization of work contents, WU (time value), and WC (occurrence of work contents in cycle). Waste operations, unbalancing of synchronous line operations, irregular-for-normal cycles and/or normal working conditions are not accepted. So, it depends on elimination (but not small amounts) as the standard model is found, especially if the improvement level so far is in poor condition. That elimination for the standard model is called the standardization effect.

Step 2. Design Specifications
Consider function and classify it with regard to BF and AF. Consider functions of operations. The innovation approach is to study the BFs of the operation, to develop a completely new method of performing them without regard to the current method, and to accomplish the BF in a minimum of time. This is the approach that would lead to the development of a nailing machine to replace a conventional hammer and nail operation.

Let's remind ourselves about the example of farmers' harvesting fruit. "Separation" is a single function and there are different operations for the function of separating fruit from trees. It is important to find creative expression of functions/purposes when current methods are observed. The expression of functions is not tied to the current methods.

Step 2.1 Basic Function and AF Classification
Once IP and OP are defined, function classification (that is, BF and AF) can be performed for each respective item, such as the processes and operations of the work performed in the model. These results are filled in on operation sheet B.

BFs are activities which directly contribute to the purpose of the process in order to change from "IP" to "OP", considering any restrictions if applicable. BFs must always be performed in order to obtain the desired change of situation, IP and OP in the model. If there are no BFs, or a purpose model has not been created, BF can never be eliminated. They can, however, be simplified, have their order changed, and be combined in different ways.

BFs can be reduced by simplification, combinations, and the like. However, they can never be eliminated with the current technical solutions without changing the IP and OP in the purpose model. BFs can be eliminated only by new technical solutions, for example, by changing the product design. When developing an improved or entirely new process, you must start by considering the necessary BFs.

AFs are activities which are necessary to assist the BFs when forming a complete process. While AFs do not contribute directly to the purpose of the process, they are essential to the purpose. They are not "waste" or unnecessary activities, but can be eliminated.

It is easy to explain these two functions in the simple assembly operations of a pen. They can be analyzed as:

- get pen and cap;
- assemble pen and cap;
- inspect product; and
- place product on a table.

In this analysis, "assemble pen and cap" is the only motion which is a BF. Only this operation contributes to increasing the OP of the product, with a direct connection between the IP and OP condition. When you take this MDC away from the work area layout, motion economy soon becomes a low level of improvement that is no longer of interest. When you create new operation methods which contain only the definitions of IP and OP and a few restrictions, it is like drawing a picture on a blank sheet of paper (Table 6.5).

You may question the relationship between work measurement level and BF/AF classification. Let's take the example of writing a sentence. "Write a sentence" is BF at the measurement level of operation, but the contents are within the elementary level; all operations are not BF. Consider the functions "get a pen from table", "move pen to writing position", "write letters" and "return pen to original position". Only "write letters" is a BF at the operation level; all other operations are AF. Then "write letters" can be divided into motions such as "move pen for writing letter", which is BF; however, "move pen to next letter" is AF. Only each letter-writing movement of the motions contributes to changing a blank piece of paper (IP condition) to completed sentences on the paper (OP). This means small levels of measurement may define BF rather than a large level of measurement. The size of level considers the efficiency of MDC activity. A small level requires more observation time than a larger one.

Returning to operation format B, various specifications may be established for production. Operation format B will be used to record not current conditions but rather the specifications which will be established as a reference to highlight the change from the current conditions. It is not necessary to consider all these items of specifications, since the desired results of method design will be established as design specification detail in the future. These will be the design requirements stated as objectives of the person who is in charge of the methods design. Operation sheet B also answers the question of how the design specification is to be

Table 6.5 Example: pen and cap assembling

Element	BF	AF
Get pen		X
Get cap		X
Assembling pen and cap	X	
Put assembled pen/cap		X

determined. The most desirable case would be for a top-down approach, with management establishing a policy that a design target is extremely important and making the MDC different from the conventional work improvement approach. It is not concerned with addressing what to do, what to aim for, or what to achieve, or with attempting to improve on past accomplishments.

In operation format A, the labor power value is calculated by dividing (WU × WC = Labor power) by ICT. This conversion is performed for the labor power reduction purpose of the MDC application. The ICT as used here stands for production speed necessary to maintain the required work or OP. If the ICT is ignored, it is possible that the production volume will drop along with any reduction in labor power.

Step 2.2 Considering Restrictions for Designing New Methods
The final aspect of the MDC specifications to be discussed is concerned with restrictions. These are not defined above, but they must be included in the design of new methods. A representative example of this is shown at the bottom of operation sheet A. Through the new method, changes in cycle time, cost, and OP conditions are achieved.

This differs according to the level of mechanization and automation stated as a condition or design specification when designing a new method. Since the MDC is not merely aimed at the conversion of work contents into a mechanical process, the desirable method keeps capital investment to a minimum.

When new production processes and manufacturing methods are designed, the cost allowed by the MDC specifications should be determined as a design condition. The ultimate selling price and manufacturing cost must be determined beforehand, bearing in mind such factors as the product's application. Since such an approach is developed in the MDC, it can be shown that the MDC provides outstanding results at a very low cost. To follow the MDC specifications is an important part of design, and through this, industrial engineers can maintain improvements at relatively low cost.

In general and practical restrictions there are three items for implementing new methods. The first is required expenses, the second is necessary terms to implement, and the third is no big change of product design. The last one is a significant issue if you want to successfully implement the new methods because any product design change of specification is not easy and takes time to do.

Design restrictions are not defined so far, but they must be included in the new methods design. Through this new method, we can see the changes in cycle time, cost, and OP conditions, but there should be restrictions for the three items of implementing new methods.

MDC does not simply direct the conversion of work contents into a mechanical process, because those improvements of methods will demand larger cost or capital investment is required. MDC can provide outstanding results at a low cost.

To follow the MDC specifications is an important part of design, and then industrial engineers can maintain improvements at a relatively low cost.

Step 3 Searching for New Methods

Step 3.1 Design Target/Kaizenshiro

The purpose of the MDC approach is not to reform or improve existing conditions. Rather, a creative approach is taken to design a new manufacturing process. This means that step 3 represents the first step of the actual MDC process. The MDC specifications are shown in operation sheet B.

Let's go back to the pen assembly example for Kaizenshiro. Assume each element requires 1 s. This means:

- get pen and cap: 1 s;
- assemble pen and cap: 1 s;
- inspect product: 1 s;
- place product on a table: 1 s.

Only the second element, "Assemble pen and cap", is the BF's element. Kaizenshiro = total current cycle time – BF element time. So Kaizenshiro in this example is 4 s – 1 s = 3 s. Kaizenshiro is the potential for improvement according to the BF of the MDC approach.

The idea of Kaizenshiro was taken from machine tool work. The amount of scrap metal produced differs according to the original size of materials and specified finishing size. The specified finishing size is fixed and the supplied materials must be cut to the specifications regardless of their original size. This means that materials thicker than necessary will require more be removed, while thinner materials require less to be removed. For the explanation of this concept, the size of the original material is the current cycle time or personnel assigned to the task and the finished size is the BF. Within the MDC the BF is not intended to be improved. This means that the amount of scrap and Kaizenshiro are not exactly the same, but since no other practical concept is available, the Kaizenshiro can be regarded as the scrap; that is, the Kaizenshiro equals the current work volume (cycle time times number of workers) minus the work volume of the BF. What changes is the amount of scrap, in accordance with the quantity of the original stock. More scrap results in higher cost, and *vice versa*. This point of explanation also applies to the MDC. It is the Kaizenshiro (Figure 6.15).

Let's introduce another practical example where the MDC is applied for improvement of the labor force on an operation. In one case, there are five workers

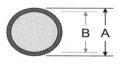

A: raw material size – present methods

B: finished material size – BF work

Kaizenshiro
= A – B
= present method work – BF work

Figure 6.15 Kaizenshiro

Figure 6.16   Ideas: fragile eggs

now and the BF of the total number of workers is 2.4. This means that 2.6 (5 – 2.4 = 2.6) is the Kaizenshiro in this example.

Accordingly, the most efficient work process is considered to be that which brings material to the finished size using only a BF. In other words, based on the definition of the BFs, all work processes are tied to increases in OP. Of course, such conditions actually do not exist in real life, but methods design should be approached considering this as the logical design target.

Step 3.2 Brainstorming for Ideas
What is an idea? An idea is just a bird's egg. Whether the egg grows well as a bird or not depends on the warming process (Figure 6.16). This means an idea for improving new working methods grows well after treating the idea. No idea guarantees the development of new methods without the right processing of the idea. This means ideas are just ideas, and are only good for the purpose of looking for solutions. This is why any ideas that are eccentric and/or have an unrealistic point of view should always be welcomed. There is no direct way to find a good or right solution. There are many books that will help you find the easy ways or solutions, for example, books on creative thinking. Ideas are ideas, they are not solutions.

We should realize using such self-help books is not the best approach as ideas are simply just ideas, and not the solution itself. They will merely open the entrance for finding solutions, but nothing more.

6.5.1   Freedom from Three Restrictions – What is the Real Reason?

To attack the root of a problem and not just the symptoms, it is often a good idea to act like an inquisitive child and repeatedly ask the question "Why?" The why technique is based on a critical perspective and asks the question why at several levels. It usually requires three to five levels to get to the real root of a problem, the cause that has to be corrected. I would like to introduce a simple example here:

A child returned from school with wet clothes.

Level 1 Mother: Why are you wet?
         Child: It's raining.
Level 2 Mother: Why didn't you have an umbrella with you?
         Child: I didn't know it was going to rain.
Level 3 Mother: Why didn't you listen to the weather forecast this morning?
         Child: I forgot.
Level 4 Mother: Why did you forget?
         Child: I'm not in the habit of listening to the weather forecast.
Level 5 Mother: What can we do to make listening to the weather forecast a habit?

Continue asking questions until the answer is a proposal to a solution. What do you think of this conversation? The real reason the child returned with wet clothes is because he didn't listen to the weather report. This is why he was not prepared with an umbrella when he left home.

There are simple but important instructions that allow us to ask good questions in order to find real reasons. Three or more questions are recommended to solicit the most current reasons. I do not say you can always find good solutions, but at least you have to discard the attitude that it's acceptable to simply agree to the first answer. The attitude of the questioner helps to reveal real reasons as well. Conflicts might be recognized if you ask questions. Discarding fictitious restrictions is essential to finding a real reason.

6.5.2 Discarding Fictitious Restrictions

"Ideas are like fragile eggs because they hatch with careful treatment. Ideas are just starting points of concrete improvement."

Look at Figure 6.17 and find solutions for a question. This is a simple but common exercise to show fictitious restrictions. You must connect nine dots by no more than four straight lines without removing your pencil from the paper while drawing the lines. We give only 3 min, and then ask for an answer. Some people cannot solve this problem and others require a long time to do so, because they unjustifiably and probably endlessly rule out the possibility of extending the lines beyond the square formed by the dots. What is the point of finding or not finding a solution? They imagine a square based on the nine spots. But such a restriction is not set forth in the question. Three minutes is long enough to find a correct solution. According to experiences, participants who cannot find a correct solution behave as if serious thinking were not permitted, even though no such restriction was mentioned in the statement of the problem. This unjustified, undesirable ruling out of a perfectly legitimate alternative or group of alternatives is a fictitious restriction (Krick 1965).

Figure 6.17 Fictitious restriction

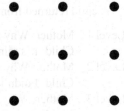

In manufacturing, our subjects are to find solutions on the working method changes, but they are not simple like this. For example, flower shops believe the design department instructs like this, the quality assurance department demands such and such. We should confirm those departments or shops believe the backgrounds are actually indicated.

The first simple answer when you ask a question to know why something is done is that that is the way it's always been done. That answer is not based on reason. Improvement ideas are hiding behind fictitious restrictions. Ideas automatically appear when you conquer fictitious restrictions.

6.5.3   Separate to Find a Solution

Look at the series of numbers in Figure 6.18 and try to memorize them. Try it silently. You may find it difficult to memorize twelve numbers in a given short time. However, some people can memorize them. How? They separate the twelve numbers into three groups of four numbers. I recommend this approach, as one who had difficulty with it at first. I am not sure what the reason is behind this, but telephone numbers are based on four digits, because humans generally find it easier to memorize it in that format. Please try this. I believe you can remember very easily four digits and continue it two times more. It is based on human experiences. There are standard time-setting techniques for mental work such as the Mento Factor, which was developed by WOFAC in the USA. The Mento Factor is a detailed predetermined time standard for mental work. It says that it is possible to remember and identify four letters and/or numbers at one time. Working methods in the shops are related to many issues like a network; this is why one solution that causes change may affect other issues. It is

Figure 6.18   One by one
for solution

981742481873

almost impossible to solve all related issues at the same time. For example, a solution is effective for reducing required man-hours but questions regarding quality and fatigue of workers remain. You should handle it case by case, and find solutions for each of them independently. Each time an improvement is made, just think about that one improvement and then move on to find solutions for any other issues.

6.5.4 Successful Brainstorming

Brainstorming (BS), is a way to find solutions. There are right ways to brainstorm. People, who were joined in a BS session asked me why it was difficult to find reasonable ideas. As advice, I would say that BS should be done correctly. For example, write all proposed ideas on sheets and stick them on the wall. Ideas are created and/or found during the time of BS. There is no need to prepare ideas by the members. An important attitude of the participants is to create new ideas through looking and understanding the ideas so far. There is no need to prepare and bring files of ideas. Coping, combining, opposite way of thinking, scale up of the other ideas, *etc.*, are expected. There are four points to follow for successful BS, those are:

- Any comments on proposed ideas are prohibited.
- A free and "run riot" climate has to be maintained.
- Quantity is more important than quality when it comes to ideas; a large number is a way to find good ideas.
- Combining and changing or improving other ideas are ways to get more ideas (Osborn 1957).

To stimulate ideas, other participants' ideas are important issues. Participants should search for ideas, not concrete improvements; this is why combinations, separations, and mixtures of previous ideas are helpful. For this purpose, any ideas should be written on flip charts, where it is easy for everyone to see them. A person writing down ideas on a piece of paper should be avoided.

6.5.5 Limited Hours of Brainstorming/Three Rounds

Long BS sessions might be tiring for participants; this is why one session of BS is recommended to last less than 60 min. If during a BS session the participants find themselves out of ideas, the session should be adjourned. Participants will find plenty of new ideas at the next session a few days later.

6.5.6 Two Stages for Identifying Ideas

BS can be divided into a green stage and a red stage. The green stage is where ideas are generated. No criticism or evaluation of the ideas may take place in this stage. The red stage is the evaluation phase. Here, the ideas generated are critically evaluated. This evaluation is not based on who came up with the ideas but on how well these ideas agree with the purpose of the productivity process. Different criteria are used to evaluate the ideas. The most important is cost per minute saved. Ideas with a low cost per minute saved are more attractive than those with a high cost per minute saved (Helmrich 2003).

6.5.7 Reasonable Theme Setting

"Reducing cycle time" and "reducing number of workers" are not suitable themes for BS to get effective ideas. Participants would not have a clear understanding about the theme because it is too wide a definition to do so. For example, if "reducing cycle time" is set as the BS theme, one participant thinks about mechanization, another about finding ideas of application of principles of motion economy, another about reducing handling distance, for example. It is desirable to get enough ideas, and those examples might result in quite a few ideas when classified into idea categories. However, they are not devoted to a limited solution subject. To avoid this situation, BS themes should be as specific as possible. As an example, I would say, "to shorten handling distance", or "to eliminate difficult adjustment work for setting jigs".

6.5.8 Demand 100 Ideas

You will get enough ideas within a short time at the beginning of a BS session. But those ideas are often just a list of ideas that are already very familiar to you, that is, there are no creative new ideas in the beginning stage of a BS session.

Those first ideas are normally simple solutions that improve or eliminate waste which has little effect on the solutions you're looking for. Plenty of ideas are there but few have an improvement effect. To avoid such an unproductive BS session, a target number of ideas is important. I recommend the target number to be 100 ideas. At the beginning stage ideas are simple and ineffective, and the first 50 or 60 ideas will be easy to come up with. However, it is more difficult to find the next 20 or 30 ideas, and the last 10 ideas are the most difficult to find. There is a high possibility of truly innovative ideas in the last 10 or 20 ideas. Those are real ideas which can be expected to lead to creative new results. Creative ideas are possible to achieve only when you perform BS correctly (Figure 6.19).

Figure 6.19 Round up leads innovative ideas

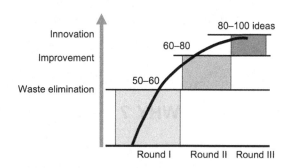

6.5.9 Aid for Finding Ideas

There is no way to find ideas, but two categories of principles are used to aid in finding ideas. Those are the four principles of improvement and the principles of motion economy: E, C, R, and S.

Ask questions about current work contents which have been studied as current methods. Describing the function is helpful to find new methods to meet the functions. The practical approach of this method of questioning is to start from the top and go down the list one by one. That is why the question "why current What" is necessary at the first BS session. Then, after finding an answer to that question, you must ask yourself if the answer eliminates the "What" and if so, what would happen? Operations of BF, BF cannot be eliminated, because elimination of BF means elimination of the design methods/module themselves. The second step is to look for C and/or R ideas regarding Where, Who, and When of Why.

The C and R ideas cannot eliminate 3 W functions or methods but partial elimination can occur. There might be operations due to operation series, so combine plural operations and functions or change 3 W to R. The final step is S, simplify the current method. Mechanization and automation belong to S; it requires expenditure. This is why How-Simplification ideas of improvements can be said to be the lowest grade of ideas. Elimination ideas normally require no expenditure, as opposed to Simplification ideas (Figure 6.20).

Another aid for searching for ideas is utilizing check lists. The most useful check list is the principle of motion economy (see Appendix 4, Motion economy). This is a fundamental tool of industrial engineers; they remember them and ask questions about model methods. A person who has the ability to remember the principles, ask questions, and find effective solutions is called a motion-minded person.

Figure 6.21 shows results of an MDC application regarding distribution of BF and AF, comparing before and after MDC. It shows a large portion of reducing production time as improved, and 29% in Figure 6.21 is the work contents, which cannot be approved as a new standard method; probably it is waste. Those were just standardization of work contents as new methods whether the worker/FM accepts them or not as standard methods. Time values are results of a 25% selection for allowed time from time study data, and frequency WU and WC. Before

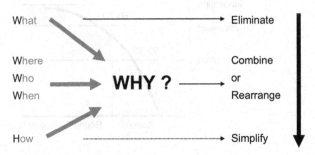

5W&1H WHY questions and improvement principles

Figure 6.20 Four principles of improvement

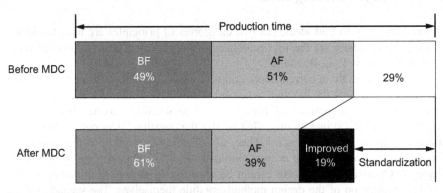

Figure 6.21 Changing BF/AF distribution on MDC results

standardization of work methods, those work contents were not controlled well as a standard method; workers had just done it their own way based on their experiences. As designed new methods, the improvement result is ratio of BF, increased from 49 to 61%.

Figure 6.22 shows the result of the improvement effect classified with the four principles of improvement: E, C, R, and S. This case is a typical result of MDC implementation. This company had seven opportunities to design. About 25% of the improvement effect belongs to E, eliminating purpose of operation; more than half of improvement belongs to the S principle of improvement, which is the lowest priority of improvement as it requires expenditure to improve. S generally requires expenditure or investment, so care should be taken to make the implementing cost as cheap as possible. Figure 6.23 shows reduced number of workers of seven proposal opportunity and each opportunity had about five modules. Total 124 workers for 47 modules were reduced as the MDC results in company C.

Step 3.3 Design New Methods and Check Targets
The design steps of the MDC are divided into fundamental design and detailed design. Design is a process in which ideas are presented and design targets achieved by incorporating them into the design.

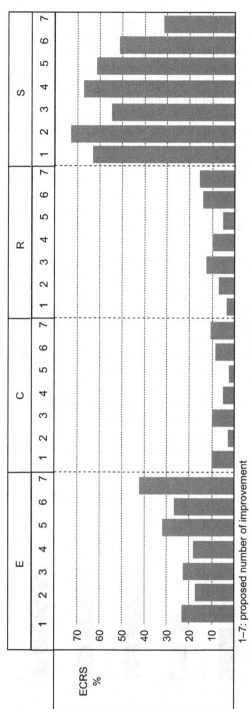

Figure 6.22 ECRS contribution for improved methods in company C

The fundamental design is the stage at which the basic details are designed using only the BFs. Since the AFs that accompany the BFs also exist, the existence of AFs and the weight of their influence differ according to the BF. Therefore, at this early stage, only the design specifications of the BFs are established.

It can be said that an infinite number of ideas are possible for the new methods, precisely reflecting the fact that the potential for improvement is also unlimited. Limits do, however, actually exist in accordance with the MDC restrictions mentioned earlier. This means that even if the idea appears to be good, it may be too expensive to implement, and experiments, or some other confirmation, may be required. Moreover, the MDC is aimed not at repeatedly producing improvement ideas but rather at establishing local targets and producing ideas that are sure to be well within the realm of possibility. This roughly resembles a combination of work simplification and changing methods, but the difference lies in the fact that ideas are generated and developed while confirming their objective and the extent of the achievement of the objective.

In the beginning of the pursuit for ideas, the objective of the processes and operations of the BFs within the model established by operation sheet A must be recognized. What – Why? Why is something performed? What would happen if it were discontinued? Application of the BS process may be used here to search for new ideas. Can we eliminate the target? If we can, the work method itself disappears and is therefore the most effective of ideas.

Elimination and simplification principles of methods improvement should not be tried in order to find improvement ideas during BS because, as described before, BFs are functions which directly connect IP conditions and OP conditions. This means that BF elimination is impossible. The principle of simplification is also not important in the fundamental design stage because there are no limited ideas. One example would be automated methods where such automated ideas may or may not be an improvement compared with the Kaizenshiro as a design target. This is why the simplification principle should not be picked up in the BS of fundamental design.

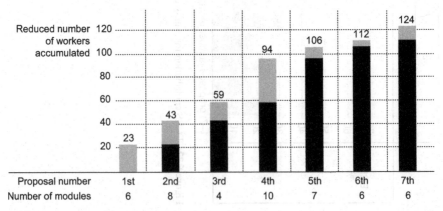

Figure 6.23   MDC activities in company C

The first step is to challenge the BFs with the questioning technique and creativity. The reason for concentrating on the BF is that we have noticed that method improvement itself is not sufficient, and methods development is necessary. Thus we have the possibility of starting on a new and higher platform. The strategy behind methods development in the MDC is built around the "black box", the principle of the purpose model. Instead of improving the existing methods, take a step backward and begin with your purpose and a clean sheet of paper.

When you have found or designed a new alternative basic process, it is time to take into account the necessary AFs. We can eliminate AFs from the old method that are no longer needed. This way of working can be compared with zero-base budgeting. You start at an absolute minimum, the core part, and then dress it up with whatever is needed, but no more. At this stage, we concentrate on elimination since AFs often can be eliminated. Using the total questioning technique and creative thinking, you can develop the new process including both BFs and AFs.

Working with the questioning technique is analytical but exacting work. Openness and tenacity are needed, and you must not give it up until you have found the best possibilities for eliminating, combining, rearranging sequences, and simplifying. If your new solution does not lead to the target set by the design specification, you must again review the method with the questioning technique and creativity to develop an even better proposal. If, despite several attempts to make improvements, the method still cannot achieve the target, the current technical solutions must be re-evaluated. We have to return to the process model in step 1.2 with IP and OP descriptions, in order to continue.

Check the design target. Design based on engineering a target set at a previous step has to be reconfirmed; it is target-oriented design. A significant point of detailed design is finding creative ideas to meet Kaizenshiro, such as a design target. This step of BS should be continued until the total amount of created ideas meet the Kaizenshiro. This is very important in the creation of effective methods design. BS should be developed not only to find ideas but also to exactly meet Kaizenshiro, such as a design target.

In practice, it might be difficult to meet the full potential of Kaizenshiro because of unexpected practical limitations, such as the cost of implementation, excessive implementation time, and engineering problems, but at least 80% of the accomplishment of the original amount of Kaizenshiro should be possible. It is normally possible to reach full or go percent of Kaizenshiro, but it is important that the target-oriented ideas are created through BS rather than traditional analytical approaches.

Methods change or improvement often is said to have no limitations. It is true from one point of view, but there is no need to always be attempting limitless, endless improvement. Management has problems to solve from time to time; this why industrial engineers are expected to meet management requirements when setting a target, that's all. Continuous attempts at improvement is a significant matter; it can go too far and might require unnecessary expenditure. This approach to improvement is not reasonable.

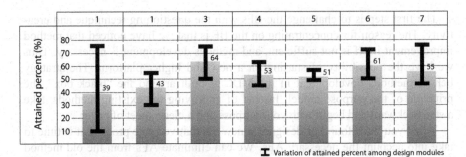

Figure 6.24   Attained percent for Kaizenshiro

Figure 6.24 shows attained results for a set improvement target as Kaizenshiro. The average attained percentage is 50 % and there is wide variation among design modules.

Step 3.4 Setting Designed New Models
As the design is realized, one new working method is developed. This is, however, a model that does not include any detailed working methods. You can find the description models in "Step 1.3 Setting Current Model". The result is a model for an improvement method. It means the model does not include any issues regarding implementation. Model work contents plus reasonable management actions lead to better practical implementation and results.

Step 4. Implementation

Step 4.1 Implementation of New Methods
The last step is implementation of the new methods as a model. The small differences between the model and the actual method that is fully implemented and/or directed should be managed by shop floor supervisors. To implement the newly developed methods successfully, designing management systems is a key factor. Good supervision, instruction by supervisors, and fine tuning new methods are also significant to success. During the early weeks and months of the implementation of a new production model, worker performance will be low and will fluctuate widely. Considerable control will need to be implemented by both supervisors and industrial engineers to monitor and improve the system.

The content that is given as a new method is a model; that means reasonable actions are expected by supervisors, support staff, and worker as well. The models which are written on paper can express explicit knowledge, but there is plenty of additional knowledge, that is, tacit knowledge.

Step 4.2 Follow-up of Implemented New Methods
There are a few practical points to implementing designed new models. Tight follow-up with support staff is expected; the new method has to be taught to each worker, one by one.

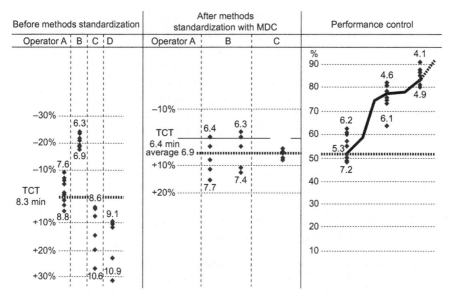

Figure 6.25 Three kinds of operation cycle time

The most common misunderstanding at the follow-up step is to mix method dimensions M and P. The minimum requirement at the beginning is perfect implementation of the proposed method; time value for operation or production volume are not to be considered at this point. This is why performance control is recommended.

Figure 6.25 is a case regarding time value before MDC and the MDC new methods implementation and performance control with engineered standard time. The FM and workers said it was difficult to implement the new designed methods due to cycle time. There was a misunderstanding of the new improved methods, which was designed with MDC. This is an example of mixing points of view regarding M and P dimensions. The manufacturing section said that the new method practice had not been completed even after it had been tried for a few days because the proposed cycle time could not be achieved. This was why the new proposed method could not be accepted as a new method. Their comment was that the allocated number of workers was OK but that the proposed cycle time could not be met due to an ineffective proposed method. There was a fundamental misunderstanding that the cycle time variance (B) was ±10%. Then, an industrial engineer checked the cycle time data before the MDC designed method; those data were collected for modeling the current methods with DTS, direct time study. They (C) fluctuated more than 10% of (B).

A point to understand at that point was what the method dimension contents were. There were a number of workers, then a new method proposed standardized work contents for each worker. Cycle time values are eventually required to be met, but they were not important as the first priority for implementing follow-up issues. What happened on standard time setting? Actual cycle time (A) variance after performance control were narrow than (B) and (C). The cycle times (A) were the results of workers being instructed and supervised well by the foreman and industrial engineers. Cycle time results under performance control were marked not only by narrow variance but also a shorter average cycle time. To improve cycle time variance, performance control is required. These better results were achieved as a result of performance control. These are normal conditions of cycle time variance; this is why performance control is important.

References

Helmrich K (2003) Productivity process: Methods and experiences of measuring and improving. International MTM Directorate, Stockholm, Sweden
Krick EV (1965) An introduction to engineering & engineering design. Wiley, New York
Osborn AF (1957) Applied imagination. Charles Scribner's and Sons, USA
Rubinstein MF (1975) Patterns of problem solving. Prentice-Hall, Englewood Cliffs, NY
Sakamoto S (1992) A practical manual of MDC. Japan Management Association, Tokyo, Japan

Chapter 7
Work Measurement

Chapter 7 demonstrates that there is no engineering without measurement. Objective judgment is not possible (whether productivity is improved or not) if it is not measured with the theoretical background of engineered standard time. This category is comprised of high-task and low-task standards, or working pace standards overall. It is possible to divide working pace into elements of skill and effort. Workers' skills do not change when simply executing jobs for a few months; however, effort does fluctuate. Workers' performance or pace fluctuates due to variation of workers' effort rather than their skill level. A worker's precise control of his own shop, for instance, will indeed improve the worker's performance.

There is no engineering without measurement, not only industrial engineering but also any engineering such as mechanical engineering, chemical engineering, electrical engineering, and so on because there is no way for objective judgment about whether productivity is improved or not if it is not measured with a theoretical background of engineered standard time.

7.1 Standard Time

7.1.1 Definition of Standard Time

Often designed methods are guaranteed regarding its prospected effectiveness without the benefit of a work measurement system. Practical work measurement systems are based on engineered time standards which are common in the world; however, they are not commonly put into practice. The usefulness of the application of work measurement to productivity improvement is often missing because management ignores the situation. There is great potential for productivity improvement with work measurement and P-control. Standard time (ST) is defined as:

- using a given method and equipment;
- under given conditions;

Figure 7.1 ST contents

> Standard time = basic time + allowance time
>   = basic time (1 + allowance %)

- by a worker with sufficient skills to do the job properly;
- by a worker who is as physically fit for the job, after adjustment to it, as the average person who can be expected to be put on the job;
- working at the pace of an approved pace standard based on the world-wide approved standard that allowed time for one unit of production. (This definition is defined by the author with reference of Mundel 1978.)

There are key words of the ST definition; those are "a. given methods, b. processing by qualified worker, c. high task/pace standard, and d. one unit of production".

This definition means ST is based on standard methods, the methods possible to follow by a qualified worker. Working pace is defined as the approved pace standard and just one unit of production. ST cannot be met unless the worker is sufficiently skilled. One unit means it is not one lot/batch, a day's standard, just one unit such as one unit of assembly, one piece of part processing, for example. Task standard is the allowable low task or high task depending on the agreement made between company management and the labor union; high task is recommended if possible. Both task standards have theoretical problems as a performance measurement of workers; it just depends on acceptance between management and the labor union.

ST contents are basic time for basic work contents and time for required and/or acceptable work contents and time for personal required time as an allowance (Figure 7.1). The relationship between ST and a day or shift working hour is shown in Figure 7.2. ST is set for one cycle base, so total hours of accumulated number of production cycles and accepted nonworking hours become a day or shift working hour. This is a key reason to measure worker performance. The difference between total hours of accumulated number of production cycles and accumulated ST for the production is the worker's performance.

Figure 6.25 from Chapter 6 shows time study results of an operation cycle time. There are three time study results. Before methods standardization, the cycle time had a lot of fluctuation compared to the target cycle time (TCT). As a result of MDC, that is after methods standardization, TCT is reduced and the fluctuation range is reduced. Then, ST was set and a P-control system implemented. The operation cycle time was reduced to much less than what it was before P-control. A key to these good results is the FM's good instructions and supervision. An average of 6.9 min is 53% of performance compared to ST.

Therefore, we should pursue the high level and make the resulting figure the ST. Also, the ST, which is set in the above-mentioned way, must be the one that is possible to reproduce completely at the actual manufacturing worksite. Using standard work methods should make it easy to reproduce. However, the work pace element of ST is the exception; it is not expected to be reproduced. In other words,

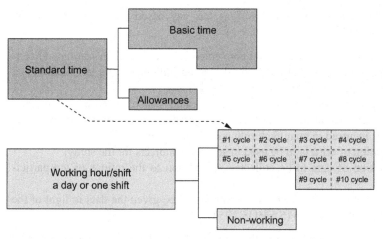

Figure 7.2 ST and shift hours (Sakamoto 1983)

we never expect an improvement of the P-level due to the operators' own efforts. The current worksite status is the way it *actually is*; improvements with tooling and machinery may be the way that you *desire*, but the way it *should be* is totally different from both of those situations when it comes to improving performance.

7.1.2 Why Standard Time Is Effective?

There is no way to manage productivity precisely without measurement by engineered standard time, which is that based on worldwide common standards such as MTM. Whether it's good or not, a method should not be measured without objective standards. There is a traditional way to evaluate methods effectiveness by comparing actual time comparisons, that is, comparing two methods of measuring time of those two practical operations. But it is incorrect because the time results for methods include and accept workers' performance results. The comparison of the effectiveness of an operation method between current and improved conditions is only possible to measure without the worker performance effect. It is measuring the difference of ST.

7.1.3 Two Standards of Working Pace

One of the most important issues of work measurement is standard of working pace. Historically, the Society for Advancement of Management in the US (SAMUS) defined "a fair day's work (AFDW)" as fair pay based on a fair day's work. SAMUS embodied:

- selection of a medium that would conveniently insure standardization and control in the methodology of obtaining the rating judgments of a large number of time study experts in many sections of the US;
- selection and presentation of tasks to be rated which, in type and number, would be representative of the repetitive operations and motion patterns most frequently incurred in manufacturing and clerical work;
- determination of criteria to be used in selecting subjects for the study;
- determination of data to be collected in addition to the ratings of the participants; and
- determination of the mathematical treatment to be given the data in light of the objectives of the research (SAMUS 1954).

We can use the SAMUS rating films for training industrial engineers including training about standard pace. Working pace distribution means a realistic working pace on shop floors is between 60 and 130%, based on a high task standard pace (Figure 7.3).

According to the principal rating scales, a 100–133% scale of 100 is a steady, deliberate, unhurried performance, as of a worker not on piece work but under proper supervision: it looks slow, but time is not being intentionally wasted while under observation, and it is 75% of 0–100% standard. On a 0–100 scale, a 100% on a standard rating scale is a brisk, business-like performance, as of an average qualified worker on piece work; necessary standard of quality and accuracy achieved with confidence, the 100 is 133 on a 100–133 scale. Those are named high task and low task standards.

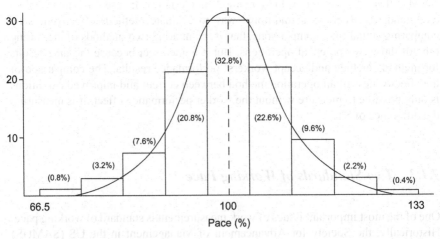

Figure 7.3 Distribution of working pace

High task means the performance of an average experienced operator working at an efficient pace, over an 8-h day under incentive conditions, without undue cumulative fatigue. It is often stated as a percentage above normal performance.

Low task is a term used to indicate that the performance rating or production standards are based on daywork levels as contrasted to high task or incentive work performance. It is sometimes taken to mean a level of performance below the level expected under measured day work (MDW) conditions (AIIE 1983).

Normal performance is defined as the following: the work output of a qualified employee which is considered acceptable in relation to standards and/or pay level, which result from a labor agreement, with or without measurement, by management or between management and the workers or their representatives.

It is an acceptable amount of work produced by a qualified employee following prescribed methods under standard conditions with an effort that does not incur cumulative fatigue from day to day (SAMUS 1954).

The MTM data system, for example, is based on a low task standard. The performance rating system used in its generation relies on a concept of normal pace that matches daywork conditions (Bayha and Karger 1977).

Work Factor Select Time is defined as that required for the average experienced operator working with good skill and effort (commensurate with physical and mental well-being) and under standard working conditions to perform one work cycle, or operation, on one unit, or piece, according to prescribed method and specified quality. The work factor select time includes no allowance for personal needs, fatigue, environmental unavoidable delays, or incentive payment.

Work Factor Select Time is not compatible with times referred to as *normal*, daywork performance, 60-min hour performance, or other terms used to indicate the work pace expected of the average worker who performs without incentive or at a level of productivity commensurate with "base rate" output.

Adequate Task Intelligence embraces the concept that the average experienced operator has sufficient intelligence to perform the task involved at a rate commensurate with Work Factor Select Time. It is expected that the subject, intent, procedures, tools, reading material, and all other items related to that task are within the operator's understanding and experiences (Quick, Duncan, and Malcolm 1962).

As a conclusion, high task standard is recommended for a company pace standard whether a company has any incentive systems or not. Don't misunderstand, high and low just means two possible standard paces. A high task standard is recommended as a company's standard for a world-wide competition situation, but they are both acceptable paces. The words "high" and "low" are just terms to distinguish between two pace standards. The family of MTM techniques are common today as work measurement techniques. A high task correction/modification coefficient set as 80% (0.80) will change the pace standard of MTM (low task) to a high task standard. This does not mean a severe standard for workers, because whether the pace standard is high task or low task, the actual P-level is up to negotiations between management and the labor union as the workers' representative. A reasonable standard percent of a company's rate is determined by the rating percent and the expected attainment. The definition of expected attainment (expected on right rating scale) is the average of qualified incentive operations on an 8-h day basis (SAMUS 1954). According to SAMUS rating films, standard values are rated by using an expected attainment.

Companies who are supported by professional consultants of work measurement. examine pace difference between WF and MTM. In Japan, the results showed the based performance standards are low task for MTM and high task for WF. The difference of working pace is 20–30% (average 25%), so the standard pace difference is recognized as 100% of MTM, 80% of WF, *i.e.*, 100% of WF is 125% of MTM. These days, WF is no longer implemented for standard setting in Japan. So, the pace of MTM standards are transferred to high task standard for companies who had previously adopted WF. The low task standard of MTM is changed to high task standard multiplied by 80% (Figure 7.4). The definition of WF is a definition of high task standard itself. The incentive system is not common in Japan today but it is common to use incentive pace for working pace; this means Japanese companies are interested beyond WF Selected Time or the high task standard definition whether they use an incentive system or not. Consultants recommend a high task standard in Japan even for those who's standard system is MTM based on a low task pace of standard. The reason is simple: the high task standard is a higher pace than the low task standard and the difference is about 25%. The difference is important when competing with other companies. It means if a company adopts a low task standard there will be a 25% loss of pace and labor cost. There is argument that says there is no difference between companies who adopt a high task standard and those who adopt a low task standard if those companies set target performance as 100% or 125%. It is not wrong from a theoretical point of view, but practices in Japan do not accept it. The reason a company im-

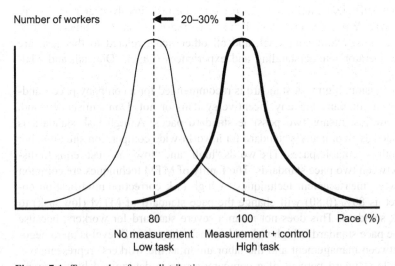

Figure 7.4 Two peaks of pace distribution

plements work measurement is the company wants to improve the low level of labors' performance, which is due to ignorance about existing standards such as the world-wide standard of working pace. One hundred percent as a target level of an average worker's performance is understood by management, supervisors, and workers to be a good target. This is why adjusting the pace difference between two PTS systems to 100% is a target.

According to the SAMUS research, there are two peaks of working pace distribution and the difference is about 25%. The difference between two pace standards' is about 20–30% (25% on average). The relation of those two paces is clearly separated as two peaks of normal distribution. Commonly, one pace is workers' with incentive and another is without incentive.

There are two common predetermined time systems: MTM and WF. It was very common to use WF for ST setting rather than MTM some 15 years ago in Japan. Despite some good experiences with WF in Japan, WF distribution has declined and MTM has become the common practice for PTS because MTM developed new systems such as MTM-2, MTM-3, MTM-V, Sequential activity and method (SAM) and so on. Also, WF could not develop any convenient and/or simple systems for ST setting practices.

Therefore, for PTS systems distribution today in Japan, MTM is the most popular one; however, the task/pace standard of MTM is equal to the low task standard regarding its definition and comparison of analysis results.

The working pace is dependent on the worker's effort. What is working pace? It is possible to divide it into two elements: skill and effort. A worker's skills do not change when he or she does a job just for a few months, but effort fluctuates from time to time. Skill level is a kind of constant due to the worker's experience with certain work, but effort level always fluctuates whether the worker is skilled or not. Let's look at Figure 7.5, which shows distribution of workers' pace for

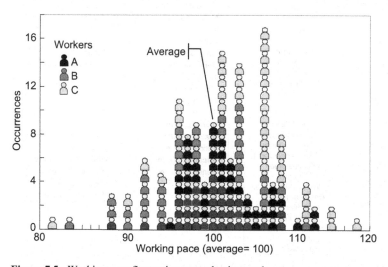

Figure 7.5 Working pace fluctuation on packaging work

packaging work. Worker C is a veteran, which means he is a well-skilled worker. Worker A and worker B are average. Given the measured results, it is not easy to understand if the worker C has the highest level of performance or not, because the highest *and* lowest paces on the work of a day belong to worker C. This result illustrates that working pace depends not only on skill but also effort. This fluctuation of working pace is not an unusual occurrence.

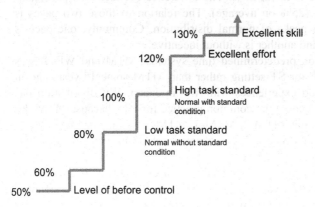

Figure 7.6 Typical level of performance

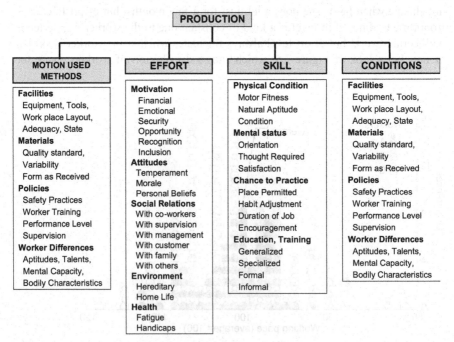

Figure 7.7 Performance contents (Bayha and Karger 1977)

Later in this book, we discuss effectiveness of P-control based on engineered standards and how the control results indicate there is plenty of room for productivity improvement. Skill is defined as the proficiency with which an operator can follow a given method. Effort is defined as the "will to work, the drive or impetus behind the motion of the workers" (Bayha and Karger 1977).

Managers believe their veteran workers always keep a higher working pace, but that is not true. Management should understand the shop floor reality. Workers' performance or pace fluctuations are always due to variation of workers' effort rather than their skill level. This means that there is plenty of room for improvement of workers' P-level by controlling or keeping a reasonable effort level. It is a key objective of setting a P-control system. Finding more than a 50% variation, after disregarding workers' performance, means there is a possibility of a more than 50% improvement in productivity.

One reason is workers' performance always fluctuates and the resulting differences are not inconsequential. Secondly, the fluctuation is not due to workers' skill, but rather workers' effort level. If workers' performance is lower than 60%, good management skills and P-control can increase that rate to 100% or more.

Typical P-levels and their backgrounds are illustrated in Figure 7.6 and Figure 7.7 introduces four reasons why fluctuation of workers can affect production (Bayha and Karger 1977).

7.1.4 How to Set Standard Time, Measuring Methods?

The first MTM system, MTM-1, was developed in 1941. Since then, the MTM family has been expanded from time to time to develop more precise and easier-to-set STs. MTM-1 and SAM data cards show those developed contents.

The basic system, MTM-1, provides a very detailed description of the method and is accordingly very method aggressive. The disadvantage is that it takes a long time to perform an analysis of a work cycle using MTM-1. In 1965, in order to widen the scope of MTM, the International MTM Directorate (IMD) developed a simplified system, MTM-2. The speed of analysis was more than doubled but at the expense of method aggressiveness and the precision of the time. During the 1980s, this development continued under new guises. SAM was developed in 1983 by the Swedish MTM Association. SAM is based on the same database as MTM-2. The goal was both to widen the range of application by increasing the speed of analysis and to broaden the range of users to include personnel close to production and designers. The speed of analysis was increased by using a sequence of Get + Put for objects, and Get + Put + Use + Return for tools.

This sequential perspective and the inclusion of accurate elements for repetitive tool use gave considerable time savings, making the analysis roughly 10 times faster than MTM-1 while also enabling a reduction of applicator deviation. Compared to MTM-2, applicator deviation has been halved.

There are two types of random deviation in MTM systems:

One is system deviation, which comes from the grouping of variants and variables.

Another is applicator deviation, which comes from users making application errors, *i.e.*, the human factor.

Of these, applicator deviation is the most serious since it undermines confidence in the analysis and its conclusions. The sequence analysis form for SAM, with its time values, is shown below. SAM is also available in data card form for the constituent Get and Put elements.

In 1987, the German MTM Association developed the MTM system to a level equivalent to that of SAM, also treating repetitive sequences separately. The system is called universal analyzing system (UAS), and provides roughly the same speed and accuracy as SAM. The distribution of the two categories of deviation in UAS is similar to that of MTM-1 and MTM-2 respectively. This means that system deviations and applicator deviations are of the same magnitude as for SAM, where the applicator deviations constitute only 1/5 of the total deviation. The UAS is documented using data cards rather than sequential analysis forms.

Both SAM and UAS have been reviewed by IMD and received IMD's "Recognition as properly developed systems" with regard to design, quality, and training. Development of the MTM family of techniques is illustrated in Figure 7.8.

SAM and UAS are available in computer-assisted forms. Computer assistance facilitates the documentation of methods and processes but also assists calculating STs and balancing lines as well as maintaining time data when method changes are implemented.

Sequential activity and methods analysis (SAM) is the most common and useful system for standard setting today. The most important point of setting ST is

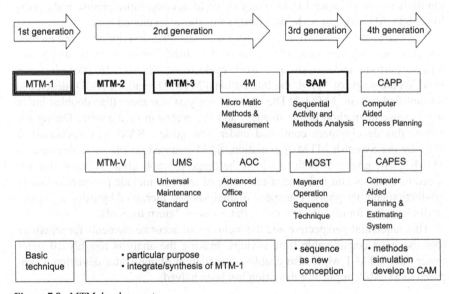

Figure 7.8 MTM development

based on a particular method. This means time is the result of methods and time value is useful to measurement as a common denominator. The full name of MTM is Methods-Time Measurement. There is a hyphen between Methods and Time, WF, or Work-Factor, is hyphenated as well, meaning that Methods or Work come first, followed by Time or Factor for finding time value. This is a very important issue regarding the setting of standards and measuring of workers' pace or performance. ST is always based on effective methods such as the one best way. The first step of setting a ST is setting standard methods.

Such an ST is useful for FM or supervisors giving instruction on shop floors. Time value is always just for measuring effectiveness; what we should understand is the methods themselves.

There are practical and useful MTM systems today, but STs are not commonly set with direct analysis of MTM systems. It is common to develop standard time data (STD), as useful ST setting tools. Developing STD requires not a small amount of human resources and time.

A practical example of the decision of whether or not to develop STD, ST data, and if so, when to develop a new STD follows.

$$I_S - I_G P \times N \left(C_{PTS} - C_{STD} \right)$$

N = Number of standards to be set/year
I_S = Initial cost to develop the STD
I_G = Initial cost for the PTS system (normally = 0)
P = Pay-off time
C_{PTS} = Cost to set a standard by direct analysis using a PTS
C_{STD} = Cost to set a standard by using an STD

(From Klaus Helmrich)

Steps to get ST for the methods "should be":

Step 1 Obtain the cooperation of the department FM.

The first step of analysis is to explain the purpose to the shop floor supervisor. No misunderstanding of the analysis should occur with a reasonable explanation to FMs first.

Step 2 Select an operator and obtain his corporation.

It is important to find standard workers who can follow work contents consistently. Unqualified workers cannot follow work contents consistently; their work contents of motions or elements fluctuate from time to time. Selecting adequate operators is the first important step to finding and defining standard operators and also standard operation contents.

It is not easy to determine standard operations when you observe nonskilled operators. In other words, you can never find gold without being at a gold mine.

Step 3 Determine whether the work is ready for study.

There are three ways of looking at production on shop floors. Those are the way it should be, the way it actually is, and the way that you wish it were. The analysis technician looks for the way it should be, but that may not be immediately apparent. To find the way it should be, analysis carefully employs techniques of basic industrial engineering such as the principles of motion economy, the attitude of motion minded, and reasonable, effective working methods. This is why "study" is recommended rather than just simple measurement and/or analysis. A few cycles of observation as prestudy are also recommended. Operators' doing jobs never use standard methods without having had the benefit of the industrial engineering point of view. Searching for standard methods means searching for the one best way.

Step 4 Obtain and record all general information about the operation and the operator.

The information includes shop name, operator name, experience, object product, drawing number, work area layout, tools, materials, *etc.*, that are useful to define objects of analysis.

Step 5 Divide the operation into elements and record a complete description of the method.

Classification of work measurement has a few levels, such as motion, element, operation, process, *etc.* Only the definition of operation is clear; operation is the smallest work contents that can be done by a worker. Element is part of operation; motion is also part of element; process is the summary of a few operations. The size of time value is not important for this classification. For example, "to reach hand to bolts bin and to take a bolt and fasten by hand and release hand from bolt", then "to take a driver and fasten bolt tightly", then set a process for these two operations. Only operation can be defined precisely even when analysis technicians are changed.

Step 6 Analysis of operation with MTM elements and calculation of a cycle time.

Then, an operation is divided into MTM elements. SAM, for example, is a line of analysis sheet for one operation. It is convenient to avoid dealing with missing MTM elements when analyzing; however, elements are a recommended level of analysis. Applying MTM elements is valuable and those regarding time value are part of the MTM analysis for getting a cycle time. Cycle means a set of work contents that can be completely repeated. It means noncyclic work contents and/or long-time cycle work contents are not included and analyzed. Those work contents are classified as allowance of delay when setting ST.

Step 7 ST calculation.

$$ST = \text{Basic/allowed time} + \text{time for allowances}$$
$$= \text{Basic/allowed time} \left(1 + \text{allowance \%}\right)$$

7.1.5 Crucial Steps for Setting Basic Time

Here are the examples of ST setting as a "should be" of basic time.

7.1.5.1 For Manual Work

MTM analysis or direct time study (DTS) should be done after preparation. A working method that you can observe at the shop is just a shop floor practice or a worker's own method. It will never become the standard. Analysis has to be done after standardization. Then the "should be standard method" is defined.

Table 7.1 shows an interesting difference between the two MTM analyses "should be" and "current actual" and DTS with stopwatch. The workers' rating value change depends on MTM analysis whether it's should be or just current analysis. Companies should always consider the "should be operation" as the standard. Several methods are done by workers when you come to shops. Column A is MTM analysis of "should be", B is the worker's own methods, which cannot

Table 7.1 Comparing time values among MTM, stopwatch and actual

No.	Operation	MTM analysis			Stopwatch (TMU)	Rating (%)
		Analysis A[1] (TMU)	Analysis B[2] (TMU)	Percentage of difference (%)		
		A	B	C = A/B	D	E
1	Door assembly	708	798	89	858	83
2	Fan unit assembly	155	252	62	325	48
3	Board assembly	258	521	50	666	39
4	Press assembly	420	468	90	608	69
5	Inner assembly	760	858	89	958	79
6	Controlbox assembly	792	966	82	1,289	61

[1] Method should be done, the one best way as a standard
[2] Worker's own way

be accepted as the standard method. C is a ratio of A and B; that is, the difference between "should be method and shouldn't be". The ratio fluctuates but the values are 50–90%. Workers' performance value percentages vary without relation to any of the workers' P-levels. ST should be always be set at the high level of study results. Otherwise you will miss evaluations regarding workers' performance. Another time value shown on the figure is DTS results. The ratio of A and D means working pace, such as rating results. The values are 39–83%. A 39% rating is an impossibly slow pace; it means not only is the worker's working pace slow, but also the worker is not following standard methods of A. Industrial engineers come to the shop to study, but a method which industrial engineers can observe at the shop has a wide range of time values, so it shouldn't be considered standard. Those works contents have to be standardized, then the range of time value decreases. There will still be variations of time value, and they have to be accepted as standardized results. The final step of such a study is selecting time value from such variations; any time values are acceptable without evaluation of pace rating. A, B, or C are all possible to set as the standard; which one is chosen to be the standard is a decision to be made by the industrial engineer.

Current working methods become an object of time study after standardization of workers' own working methods by industrial engineers. MTM analysis applies to the standardized one method. But DTS gets more than two times the value for operations. Which one is a suitable standard method? This is up to the industrial engineers to decide. This means any measured time value after method standardization is possible as ST. The first observation time values by DTS fluctuate, as shown in Figure 7.9 (Sakamoto 1983). This is where industrial engineers come in to decrease that fluctuation. The results still fluctuate but any time value is possible to adopt as a standard.

Natural fluctuations of a worker's operation never become the standard; industrial engineers find unnecessary work contents, standardizing tools materials, work area layout, and so on. It becomes a "should be standard method". There is no problem when time value is obtained with MTM, but when you use DTS, there is still variation of observed time value. A recommended time value to select is the minimum value of them.

Why did the P-level start at 50 or 60%? The answer is provided by precise analysis of the higher level of work contents; analysis without the use of MTM can never find such a higher level of a standard. This is the point of view that ST says "engineered standard time is important". There are 6 operations as examples in Table 7.1. MTM analysis has two operations, A and B. A is a method that should be done, the one best way as a standard; B is the worker's own methods. Column C is a ratio of the two methods. Operation number 3 on Table 7.1, board assembly, is 50% of the worker's own methods which, based on experience, is on average around 80% or less. Column D shows present working time measured by a stopwatch; rating values are 39–83% compared to how the method should be done, with the one best way as a standard. Don't misunderstand; workers do work at such a low level, but they do not know standard methods as a base of ST. It comes from workers' or FMs' ignorance about standard operation contents. Which one is better

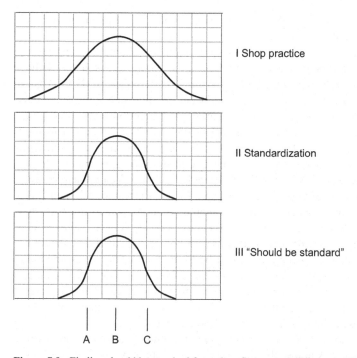

I Shop practice

II Standardization

III "Should be standard"

A   B   C

Figure 7.9 Finding should be standard from shop floor actual (Sakamoto 1983)

as a company standard? Column B is the workers' own way of production; it is not the company standard, it is things that happened at the shop before the company implemented a labor P-control system based on engineered standard time.

The differences between A as a method that should be done, the one best way as a standard, and B, the worker's own way of methods are discussed below.

The normal working area, with regard to Reach and Move distance should be as short as possible. Less than 30 cm is the best because 30 cm is a normal working area. Ralf M. Barnes describes it as follows. The normal working area for the right hand is determined by an arc drawn with a sweep of the right hand across the table. The forearm only is extended, and the upper arm hangs at the side of the body in a natural position. The upper arm tends to swing away as the hand moves toward the outer part of the work place. The normal working area for the left hand is determined in a similar manner. The normal arc drawn with the right and left hands will cross each other at a point in front of the worker. The overlapping area constitutes a zone in which two-handed work may be done most conveniently. There is a maximum working area for the right hand and for the left hand, working separately, and for both hands working together. The maximum working area for the right hand is determined by an arc drawn with a sweep of the right hand across the table, with the arm pivoted at the right shoulder. The maximum working area for the left overlapping area formed by these two maximum arcs constitutes a zone beyond which two-handed work cannot be performed without causing considerable disturbance of posture, accompanied by excessive fatigue. Normal area distance of an arm is 15.5" for males, 14" for females and the maximum is 26.5" for males, 23.5" for females (Barnes 1980).

For a method that should be done, industrial engineers should do their best to find a solution to establish a normal working area. Another reason for finding a Reach and/or Move distance for a method is the distance codes of MTM-2. They are less than 5 cm, 15 cm, 30 cm, 45 cm, and 80 cm. Many distance measurements that are beyond the distance codes are a result of the necessity to take steps, turn the body, bend, and arise. These issues affect the definition of how a method should be done, and so should be taken into consideration when setting the standard methods. The number of parts to be handled also needs to be taken into consideration when determining how a method should be done. Not only assembling, but all Reach and Move motions are considered when determining how a Simo operation that uses both hands as a method should be done. Placing finished parts or products precisely may or may not be necessary. Tossing it away is better than placing it precisely on top of other materials or a box as a method that should be done. Hanging tools such as drivers should be located above distances and hang methods should be acceptable, effortless Move and Release motions. A fixed location for tools but no requirement for handling them are components of how a method should be done. Principles of motion economy are also important when determining how a method should be done (Barnes 1980).

More considerations when determining how a method should be done as the one best way are, for example, the number of times a bolt and/or nut need to be twisted, the number of movements when cleaning up with a rag, the number of times a hand or power tool needs to be turned, the number of times pasted parts need to be rubbed, and so on. Table 7.2 shows the time difference of MTM analysis between accept present method of analysis and considering should be method of analysis. Table 7.3 details MTM analysis.

In a case of operations which contain both manual and machine processes, there are three possible relationships between the processes. They are: separately as a series, partial overlapping, and sweeping overlapping. The issues for setting standards for these Multi-Person-Machine works follow.

Table 7.2 Two MTM analysis for an operation

Two MTM analysis of an operation	
Present practice (TMU) 929	Should be (TMU) 817

Difference between two analysis		**MTM analysis/TMU difference (TMU)**
1	Methods change A	43
2	Standardize move distance	8
3	Methods change B	25
4	Standardize move distance	36
	Total of difference	112

Table 7.3 Details of Two MTM analyses

Present practice						
Door assembly	**Hand (RH/LH)**	**Analysis**				**TMU**
1 Door panel on a bench	RH	GB80	PA80		S	61
2 Position airgun	RH	GB45	PA45			33
3 Browing for take off protect sheet	RH				PT	34
4 Position another	RH		PA45			15
5 Browing for take off protect sheet	RH				PT	34
6 Return airgun	RH		PA45			15
7 Move to upright position	LH	GB45				38
8 Take off protect sheet	LH		PA45			45
9 Reach to under panel	RH	GB45				18
10 Return to horizontal position	RH		PA30			11
11 Take off left protect sheet	RH	GB30	PA45			59
12 180 turn over	RH	GB45	PA30			40
13 Put part A (2) and B (2)	RH	GB80	PA80			43
14 Part C, part D (2) get and put	RH	GC45	PB30			46
15 Put Part C, part D (2)	RH		PB30			19
16 Get tape (2) and put no. 1	RH	GB45	PB15	PB45		72
17 Rub tape	LH		PA15			6
18 Put tape (no. 2)	RH		PA15	PB45		54
19 Rub tape	LH		PA15			6
20 Get part E (2), place no. 1	RH	GB45	PC80			59
21 Place no. 2	RH		PC80			41
22 Get part F and 180 turn over, one for both hands	RH	GB30	PA45			43
23 Place F and insert	RH		PB5			10
24 Get hammer and bit	RH	GB30	PA30	PB30		44
25 Get part F (no. 2)	RH	GB80				23
26 Get door panel, 90 turn	LH	GB5	PA45			22
27 Place F (no. 2) and insert	LH	PB5				10
28 Hit F and return hammer	RH	PA30	PB15			26
29 Completed assembly to conveyer	RH	GB45	PA80		S	70
		manual time	929	process time	68	

Table 7.3 (*Continued*)

Should be method Door assembly	Hand (RH/LH)	Analysis			TMU
1 Door panel on a bench	RH	GB80	PA80	S	61
2 Position airgun	RH	GB45	PA45		33
3 Browing for take off protect sheet	RH			PT	34
4 Position another	RH		PA45		15
5 Browing for take off protect sheet	RH			PT	34
6 Return airgun	RH		PA45		15
7 Move to upright position	RH	GB45			38
8 Take off protect sheet	LH		PA45		45
9 Reach to under panel	RH	GB45			18
10 Return to horizontal position	RH		PA30		11
11 Take off left protect sheet	RH	GB30	PA45		59
12 180 turn over	RH	GB45	PA30		40
13 Eliminate					0
14 Part C, part D (2) put (reduce part C)	RH	GC30	PB30		42
15 Part C, part D (2) put	RH		PB30		19
16 Get tape (2) and put no. 1	RH		PB15		72
17 Rub tape	LH		PA15		6
18 Put tape (no. 2)	RH		PA15	PB45	54
19 Rub tape	LH		PA15		6
20 Get part E (2), place no. 1, (reduce G distance)	RH	GB30	PC80		55
21 Place no. 2	RH		PC80		41
22 Get part F and 180 turn over, one for both hands	RH	GB30	PC80		43
23 Place F and insert	RH		PB30		19
24 Get hammer and bit	RH	GB30	PB30		33
25 Eliminate					0
26 Get door panel, 90 turn	LH	GB5	PA45		22
27 Place F (no. 2) and insert	RH		PB5		10
28 Hit F and return hammer	RH	PA30	PB15		26
29 Completed assembly to conveyer (reduced distance)	RH	GB30	PA80		34
		manual time	817	process time	68

7.1.5.2 Multi-person Machine

This is a case of a cycle operation which is a series: first a manual operation then a machine operation, then a manual operation. The sum of the manual times and the machine time is the cycle time of this operation. One issue to consider is that the cycle time is the basic time of a content of ST. To calculate ST one must apply allowances. Applied allowances differ for manual time and machine processing time. For a manual operation of moving materials to a die and then removing it from the die an allowance of manual work is applied such as fatigue allowance (F), personal allowance (P), and delay allowance (D). However, for a machine processing time only D is applied. Allowances are generally 5% for each P, F, and D for machines. So, assume 0.100 and 0.20 min for the manual operations and 0.25 min for machine processing. The ST is calculated thusly:

$$(0.10 + 0.20)(1 + 0.15) + 0.2(1 + 0.05) = 0.608 \, \text{min/cycle}$$

Details of these relationships can be had by writing a chart of the multi-person machine relationship.

Partial overlapping is a situation where an operation, C, is done parallel with the part of the machine processing time (Figure 7.10) (Sakamoto 1983). There is some difficulty with a case like this. Part of the machine processing time happens as a series within the cycle. The industrial engineer has to find standard methods like this, not just accept a case like that mentioned above as "separately as a series". The industrial engineer has to make an effort to find cases of partial overlapping as much as possible. Simple acceptance of the above case increases its cycle time more than partial overlapping. Application allowances are just D for a part of a series and P, F, and D for overlapping.

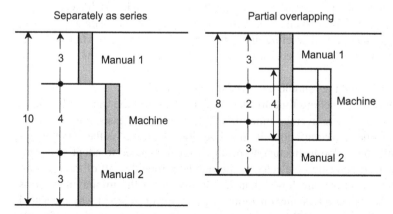

Figure 7.10 Manual and machine time relation for should be standard

Sweeping overlapping is a case where an operator is watching a machine or processing time all the time or an operator is supplying material for the machine while the machine conducts its own process. Application of allowance is just D for supplying during machine processing time. In the case where those operations are controlled by the worker, P, F, and D are applied.

7.1.5.3 For Mechanized Process Time

Facilities that process foods, raw materials, pharmaceuticals, cosmetics, and so on have to set STs. When setting machine tool work, recommended cutting conditions can be found in the machining data handbook (MDH) as a kind of world standard. But the specific purposes of any given facility are not common data in the market.

DTS that just measures current conditions is not recommended as a ST.

There are three approaches for setting those uncommon machines (Sakamoto 1983).

1. Apply theories which have world-wide approval.
2. Generate or develop theories with specially designed experimentation.
3. Find the "should be" standard through DTS.

Approach 1. Apply theories which have world-wide approval.
MDH for machine tools is an example of world-wide approved cutting conditions. Welding, punch press, and hydraulic movements can qualify for setting standards according to the world standard.

Approach 2. Generate or develop theories with specially designed experimentation. Industrial engineers should understand that any machines or large-scale facilities are controlled with mechanical and/or electrical engineering. It is relatively easy to establish the operation time for machines or operations that use electric motors and hydraulic power for movement. On the other hand, it is difficult to find a common standard for manual work such as painting by hand or with a spray gun, or for a filing operation. Therefore, industrial engineers are required to do experimentation with standardized conditions and find a rule or theory as a company inside standard.

Approach 3. Find the "should be" standard through DTS.
The last approach to setting time for the standard is to find the "should be" standard through DTS. The average time of DTS results is not reasonable as an ST. Those time values have variation; the first question should be "is the data the reason". A minimum value in fluctuating results is recommended to adopt as a standard. If the average value is selected as the standard, roughly half of the opportunities of the operations are lower than it. Meanwhile, if the minimum value is selected as the standard then any operation suggests the question of why the actual cannot meet the standard. This question regarding actual time or the difference

from the standard leads to the current level rising to a better level. Things that can be discovered through work measurement are different depending on the level or contents of the standard. This is an important point for management who are interested in improving performance or productivity.

7.1.6 Maintaining Standard Methods and Time

In order to have successful productivity regarding P and to always maintain reasonable accuracy regarding ST, it's necessary for the FM and the worker to trust the contents of the ST. The following circumstances are required in order to maintain the best ST situation.

7.1.6.1 Convenient Tools for Setting Standard Time

An MTM system like SAM is very convenient; however, direct SAM analysis can not be recommended because it requires too much time to set the ST. Instead, STD, is recommended. The development takes more than 6 months, but it takes much less time to set the ST with STD, and it's convenient. If such convenience is not sustained, ST setting or revisions are neglected as a result. That means the standard methods which workers are following become out of date. Such a situation absolutely should be avoided.

MTM analysis speed, called the measuring factor, times the nonrepetitive cycle time. The following are required to make the analysis:

- MTM-1:   350
- MTM-2:   150
- SAM: 50

For example, the ratio for MTM analysis speed for MTM-1 is 350:1. Therefore, it would require, for example, 350 min of analysis time for 1 min of nonrepetitive cycle time. The MTM-2 ratio is 150:1, *i.e.*, 150 min of analysis time for 1 min of nonrepetitive cycle time. SAM is 3 times quicker than MTM-2; its ratio is 50:1, so only 50 min of analysis time would be required for 1 min of nonrepetitive cycle time.

These analysis time requirements could be unacceptable for daily ST settings if there are many different products for which standards are to be set. This is why it is convenient to develop an STD, ST data, and then use a computer to set standards for all product variants. It is recommended that a company custom design computer software to obtain STD. There are several computer software applications for setting ST, but it is difficult and inconvenient to connect a company's custom-designed STD software to an open market computer system.

Another point to develop an STD system is balancing time; it is the time required for the STD system to balance out random deviations when building the STD. It is important to set the desired level of accuracy and build the STD system to meet this requirement. The amount of time of analysis must be accumulated when using an MTM system accurately if that system is to guarantee a specific level of confidence. The balancing time of SAM is 6,497 TMU, *i.e.*, random errors are leveled out with 95% probability at a maximum +/– 5% at 6,497 TMU, compared to an error-free MTM-1 analysis. The total balance time in relation to an error-free system is approximately 8,600 TMU. A SAM analysis of 5 min thus has a measurement accuracy which, with 95% probability, is less than +/– 5%. The corresponding time for MTM-2 is approximately 3.6 min.

The random errors in MTM systems are normally divided into System errors and Applicator errors. The most serious error is the Applicator error as that tends to make people lose confidence in the results. All official MTM systems have half of the errors coming from the System and the other half from the Applicator. SAM is in this aspect different; only one fifth of the errors come from the Applicator. In fact, SAM is more accurate than MTM-2 when only looking at Applicator error. This was one of the basic design criteria for SAM.

Simplifying an MTM system or any STD system is possible, but accuracy should be considered. There is a useful report from the Swedish Federation of Productivity Services about developing SAM. It states that:

Remaining differences between the systems are referred to as systematic errors and are, by definition, related to SAM. SAM, thus, has been tested with regard to systematic errors in relation to MTM-1 which manifests itself as a difference in the performance norm. Furthermore SAM has – as have other ST systems – certain random differences which have a tendency to equalize themselves the longer time the analysis covers. These variations are usually called system distribution". (Swedish Federation of Productivity Services 1993)

Standard operating procedures (SOP) have to be issued to the shop simultaneously with STs. SOP includes working area layout based on handling, such as reach and move, distances, location of tools, materials and pallets/containers, Simo-motion of both hands, handling number of each operation, and so on, with corresponding ST time value. The FM instructs this content to each worker when starting assigned work.

7.1.6.2 Revising Standard Time

Accurate ST is also maintained with revisions from the manufacturing division. There are several reasons that current standard methods could be unsuitable after product design changes have been made: materials specification, suppliers' quality issues, machine or facilities matter, production pace or tact, and so on. The production department needs to be made aware of every change. This is why any change requirements are proposed by the production department. Occasionally, revisions to the ST from the FM or managers are necessary to always maintain reasonably accurate ST.

7.1.6.3 Auditing Standard Time

The last matter is periodical auditing of current standards. Auditing means a comparison between current standard time using STD and the direct MTM systems analysis such as SAM or MTM-2 to make sure items at workstations are good enough. Acceptable error is defined as a +/– 10% of difference. If there is a 10% difference for each auditing item after summing up measurements of a shop unit or FM for week, then the actual difference on a weekly basis and shop unit performance are good enough; it means any unrealistic variation is not happening. It is recommended to do an audit at least every 3 months; but every month is much better.

Typical causes of any incorrect STs are found as a result of auditing are: a revision demand was missed, incorrect using of STD, a necessity to revise STD, incorrect recording to database, *etc.* The first cause, missing a current standard revision demand, is an important reason to make sure the ST is accurately measuring workers' performance. Monthly audits are recommended for unusually high or low performance at workstations compared to the average of plant performance. Tri-monthly audits are required for results of randomly sampled workstations whether P-level is high or low. The purpose of auditing is not to find processes of incorrect ST; it is to identify correct ST with regular conditions.

Figure 7.11 shows the results of ST auditing for a subassembly work. Before auditing, labor performance at the shop was 133%. It is an unusually high level of performance. As a result of auditing, SAM analysis was 1,096 TMU for the current ST (basic time); it was 1,260 TMU for the old ST before auditing. The industrial engineer set the new standard method as the "should be" base. The difference indicates that current, loose methods were changed. The content is "simultaneous motion for getting materials, same as for assembly, and reducing reach and move distance from 80 cm to 45 cm".

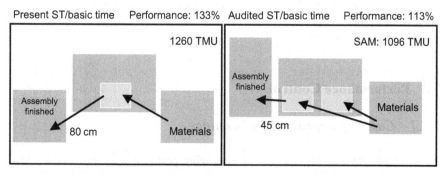

Figure 7.11 An example: Auditing result ST

7.2 Allowances

ST is "basic/allowed time + allowances". Basic/allowed time is set by MTM as a universal standard but there is no way for an allowance to be made a standard or universal. Allowance is a question of coordination between management and the labor union.

Typical allowances of ST setting are fatigue allowance (F), personal allowance (P), and delay allowance (D).

Personal allowances are for personal time, such as going to the bathroom, drinking water, smoking, *etc.*; anything that is necessary for living in a day as a human. Fatigue allowances are for taking short breaks to recover from fatigue due to doing operations or the conditions surrounding the work, such as temperature, noise, smell, and dust. This is why these two allowances are not easy to separate theoretically. And the last type of allowance is delay allowances, which are for irregular work contents and irregular occurrences; those delays are not predictable, and the necessary time value is not constant, nor is it large.

There is no universal standard for these allowances, so when making a production study or WS study there is no standard value for the allowances. However, benchmark values from industries are useful to adopt to know the actual conditions for allowances. A very rough benchmark is 5% for each allowance; a total of 15% is a common value to adopt for many companies.

For machine or processing time only the machine delay allowance is applied. It is normally less than D, delay allowances for workers. The condition for this applies only to fully automated processing machines or the facility itself. It means electric or pneumatic power tools, such as drivers and drills, are not subject to machine delay allowances.

There might be need for other allowances, for example a learning allowance; a machine interference allowance depends on the purpose of setting ST. The most common purpose of ST setting, however, is labor P-control; P, F, and D for irregular work is also common.

The contents of the delay allowance is the work itself. There are three reasons for that. They are (Sakamoto 1983):

- Occurrences are not cyclic; it is difficult to predict their occurrence.
- Time values have wide fluctuation; it is difficult to set a fixed time value.
- Time values are small.

7.3 Performance Control

7.3.1 Cases of Improved Performance

Figures 7.12 and 7.13 show cases of improving performance. The results show more than two times of labor performance improvement.

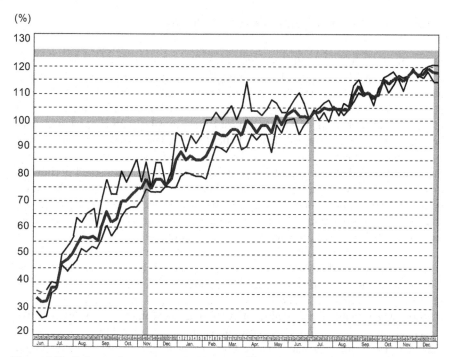

Figure 7.12 Labor performance progress: company B

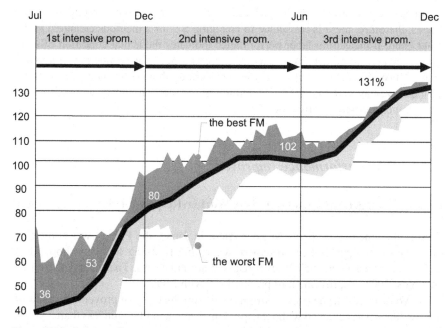

Figure 7.13 Labor performance progress: company A

7.3.2   Three Control Systems for Shop Floors

There are three typical control systems regarding workers. The first is P-control with measuring ST. The second is consumed man-hour control, which is just actual elapsed hour measuring. The third is control of the number of workers. The last one is the weakest labor control system, because those labor activity contents are not identified and controlled. The next weakest labor control system is consumed man-hour measurement and control. This is better than the last system, but management cannot take reasonable actions. These last two control systems are accepted due to the fact that it is difficult to set STs for those kinds of work and small portions in total labor. A very rough percent of distribution is 70–20–10% for each control system for shop floor work. This means that management finds solutions to set standards for consumed man-hour measure and control areas, and to measure consumed man-hours at least for control of the number of workers' areas.

There are two pay systems: incentive and the day work system. A further classification is measured labors' consumed hour or not. Then there are two systems that measure and pay, with performance measured by ST and day work pay measured by performance. The one is a wage incentive pay system and the other is a MDW system without a wage incentive. Management interest in wage incentive systems is not common today because the value of wage incentives for labor has declined. Time methods may change due to changing STs, which may reduce labor's P-level and wages, so implementing new methods are not welcome to labor and labor unions. Even if a new method is accepted by labor it is still a concern because if there is a decline of labor's performance it will directly affect his wages, therefore, there is resistance to changing ST for a methods change. On the other hand, if measuring labor performance indicates the actual condition of shop floors, then it can be determined where there is a need for productivity improvement. MDW derives a conclusion from measuring labor performance but without an incentive pay system. MDW is common in Japanese industries and contributes much to productivity improvement.

7.3.3   Why Performance Improved?

7.3.3.1   Management Fails to Manage Worker-Responsible Losses

Before implementation of the MDW system in Japan, the P-level was some 50–60%. Those companies had no experience with ST setting for their shop floors before. One year later, their P-level reached around 120%. That means there was an almost 200% improvement of productivity with MDW.

As you can see in the previous figures, the productivity improvement result of almost 200% was achieved without capital investment, and within less than 2 years. What was the situation that caused the improvement in workers' performance?

Similar kinds of performance improvement results are easy to find in Japanese manufacturers who adopt an engineered standard time that is set by using MTM for a high task standard pace. The 1-year record can be divided into three levels of improvement activity, which are explained below in Section 7.3.3.2 Three Intensive Promotion Stages Are Set.

Management still can't implement the first step of P-control which is to measure the P-level of the actual performance of designated work methods in ST. Promoting the mind innovation of management in P-control and the P-level is the barometer, which means all the sections and the managers need to fulfill their own responsibilities regarding steps that should be or can be taken.

P-control is an extremely useful and effective thing for management and it is no more and no less than that. It is something indispensable for management and real P-control lies in such a control system. Bear this idea in mind and take the right steps. For each operator's part, the FM can ensure the right operation thinking in the right way.

7.3.3.2 Three Intensive Promotion Stages Are Set

P-control is increased by 3 Intensive Promotion Stages. Each of the 3 stages is comprised of certain steps that have the goal of reaching a P-level improvement of 80%, 100%, and 120%, respectively. Before discussing each Intensive Promotion Stage individually, some background information the stages and on performance assessment in general is necessary.

Each stage is reached with a specially organized support staff. What is needed, and it is important, is to steadily work on each activity and complete them one by one. Each stage requires 4–6 months. A project proceeds with the promotion aiming for P-level at each level for the following intensive promotion stages.

The basic idea of P-level improvement is "how to work efficiently" not "how to work quickly". Also, it is not a methods improvement activity; P-level improvement never deals with operation improvement. P-level is the comparison of the actual operation with the ST and operation methods; therefore, the purpose of P-control is to bring the actual operation close to these standards.

The calculation formula of performance is:

$$\text{Performance} = (\text{produced items} \times \text{ST})/(\text{shift hours} - \text{idle or nonproductive time})$$
$$\times 100$$
$$= (\text{produced standard hours})/(\text{consumed hours for net producing})$$
$$\times 100$$

Performance does not improve completely smoothly. There are 2 stagnation weeks around 80% and 100%. The reason is performance never changes without a new level of actions. Certain actions to meet 80%, for example, do not increase performance higher than 80%. This level of performance should be maintained for 1 year and the FM and the shop workers need a few weeks or months to practice the performance. Understand and take actions regarding the previous level of performance that caused the weeks of stagnation.

Each performance improvement stage takes about 12–16 months to complete.

It has been my experience that sometimes management misses a great opportunity to improve the P-level for a significant productivity improvement. A global level manufacturer will never overlook such an opportunity to double its productivity without special capital investment. It is only common sense for management to realize the treasure they hold in their hands.

There are no special tricks to improving workers' performance. Improving workers' performance is simply a matter of bringing the level of the actual status at the shop floor to a world-wide standard of MTM as ST. The ST represents the converted time value from the standard operation methods based on the world-wide standard, and ST is specified by manning and TCT. What we can do first at the shop floors is to reach the world standard of work. Workers have to implement a number of modifications at the shop floors in order to practice the standard operation. To be more concrete, workers have to observe the specified operation methods and layout on ST on a constant, daily basis. These implementations result in P-level improvement, since P-level never improves without modifications to conform to the standard.

The first stage is to reach 80%, second stage is to reach 100%, and third stage is to reach a 120% or higher level of performance. Reaching these three levels has particular meaning. The chart shows that reaching each level takes some months and performance must stay at that level for months before the next level can be reached. Through many attempts at improving performance, the three stages depend on reaching and setting 80%, 100%, and 120%. Each stage takes around 6 months according to experience.

1st Intensive Promotion

The starting point of performance is 50–60% in plant level, some 500 more workers. Performance is not just operators' working pace; performance is not equal to working pace. One must ask why such a low level of performance was marked at the starting point of measuring performance. Before setting ST, workers and FM ignored the necessity of dividing working hours and nonworking hours. This means, they believed any of consumed time at shops was their responsibility. It is not true, however, some of contents or reasons of consumed hours are higher management's responsibility, such as lack of staff support/services, lack of materials, poor quality disturbing normal operations, any other idle time due to machines/facilities, production planning and control, quality control, manpower resources, and so on. So, the first stage of performance improvement is to concentrate and divide consumed hours into working hours and nonworking hours. P-level reaches 80% with these actions.

The first stage is to reach 80%, the second stage is 100%, and the third stage is a 120% or higher level of performance. Reaching each of these three levels has particular meaning. The chart shows it takes around 6 months to reach each level and it must stay at that level for months before new actions for the next level can take place. At the beginning of P-control, the first P-level was about 40%. The level fluctuates depending on the plant, the FM, or the measurement unit. However, the levels are more or less the same, which is about 40%, the measured result by global ST, *i.e.*, 40% against the standard of 100%. It is not too much to say that this P-level is extraordinarily low. In this measurement, the actual time is measured by the acceptable time range of ST, and this result shows that 10 h are spent for the operation, which should be completed within 4 h. Even giving favorable consideration to many underlying circumstances, this P-level result is too low and the worker is taking too much time.

One conceivable cause is the low performance of workers but it is impossible to think that the actual work pace is at the level of 40% is impossible. In other words, it is highly unlikely that the result falls to below 60% since the worker's P-level at the shop is much higher. Then, what brings down such a P-level? The sole and significant reason is that idle time has been reported as operations. As is often the case with this matter, the cause is the idle time that is not exposed. It is unknowingly hidden because some existing idle time is not recognized. This actual situation is not uncommon in the manufacturing shop floor where the production planning (production unit and time required or lead time) is given high priority without looking at other issues.

The problem with this extremely low P-level should be recognized and all the manufacturing-related managers have to take the right actions against it. The workers never do their jobs taking 10 h for 4-h work. It is no exaggeration to say that the managers make them do that, which is highly irrational. It can also be said that the managers use 10 workers for 4-worker work, whereas the workers are keeping their P-level higher than 40%. The managers should be held responsible for this issue.

Figure 7.14 shows such a condition; U percent declines from 90% to around 75% for 4 months then afterwards the U percent remains around 80%. This is why for the first 3 months a labor performance lower than 50% is not real labor performance; it includes incorrect reporting of U.

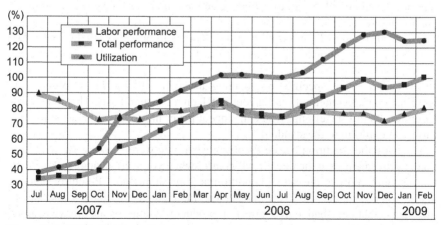

Figure 7.14 Utilization is steady

2nd Intensive Promotion: to Follow Standard Methods Exactly

The second step targets a 100% P-level; it means workers follow standard methods which are defined/set as ST. The FM provides good instruction to the workers. There are no special actions to reach 100%. The key is just to follow defined standard methods. ST includes workers' experiences, skills, and effort. The difference between actual time and ST is equal to the performance difference compared to 100%. FMs instructions to operators as to concrete methods and reasonable supervision are fundamental conditions to reach 100% as a standard P-level.

An important matter is to bring the level of the actual status at the shop floor to the global standard. It may sound difficult, but the ST represents the converted time value from the operation method based on global standards and ST is specified by manning and TCT. The reason for "Challenge Time" implementation at the shop floors by reducing manning or setting the targeted production volume per hour is to have an experience of actual operation based on ST, even if it is for a short period of time. A number of modifications are implemented at the shop floors in order to practice the standard operation, to be more concrete, observing the specified operation method and layout in ST on a constant basis. This implementation results in P-level improvement, since it never improves without the modifications to conform to the standard. Workers follow standard methods exactly with their performance-oriented mind and the target level of performance is 100%.

The steps to reach 100% of performance especially include the need to review the clear view at the shop floor, the height of workstations, partitions of work areas, parts supply, and so on. Complete implementation of these issues are basic actions to reach 100% performance. Being aware of the worker and FM's awareness of standard methods is key for the achievement of the 2nd Intensive Promotion, *i.e.*, a 100% P-level.

All the shop floors have been working on P control aiming for the achievement of 100% of P-level of the global standard. However, it is very regrettable that there are a number of different new actions ahead which should be implemented by all means in order to achieve a 100% or even higher level of excellence. Therefore, it is required to have all the effective actions completed in a steady fashion in order to move forward.

The goal of the FM's effort at the shop floors is to bring the level of the actual status at the shop floor to the global standard. It may sound difficult, but the ST represents the converted time value from the operation method based on global standards, and ST is specified by manning and TCT. The FM has to observe the specified operation methods and layout in ST on a constant basis. This implementation results in P-level improvement, since it never improves without the modifications to conform to the standard.

Compare the condition before P-control when you were working on the operation improvement with the one after P-control when you were changing the operation methods: how is the actual condition at your shop floor? P-control practice makes you realize that the lively atmosphere, the numbers of changes, and the actual results that have had a great effect on productivity improvement have resulted in a situation that is incomparably different from what it was before P-control. If you can't see the difference, look back and think with the FM. Your diligent effort at the shop floor was the challenge for the global standard and professional job performance.

3rd Intensive Promotion: a Challenge to Achieve a Possible, Higher Target Level
The third step is a challenge to achieve a reachable, higher level of performance. Reachable means it is possible, but not necessary, to reach that level. When does a reachable performance become an actual performance? The points are also possible to find in the definition of ST. ST includes some 5% of delay allowance; working pace is possibly 120% or more of ST; the skill level of workers in ST is based on the ST norm, that is not super skill. Effort is also part of ST, but effort levels of workers fluctuate without workers' and FMs' recognition. This means there are several reasons and conditions that workers can work faster compared to ST. Reachable performance is calculated theoretically.

At the beginning of the third intensive promotion, when the FM should have reached 100% of level, a seminar for the FM about P-control is implemented. In the seminar, the FM carefully watches the video of the actual performance, trying to assess and rate the work pace (generally speaking, the speed); however, they cannot make the right assessment. One of the reasons is that they judge the work pace as being slow or fast only by looking at the performance without understanding the standard of 100% or appropriate assumptions of the standard operation method and time. What is important here is that they should fully understand the guidelines of the third intensive promotion that theoretically-supported approaches are inevitable for P-level improvement. Even though the work pace changes owing to only the worker's effort, P-level never improves without exercising ingenuity in order to motivate the workers. One of the misunderstandings you have is about changing or improving the work methods. However, methods improvement does not work for P-level. In fact, it is wrong. This way of thinking reveals that you do not correctly understand the distinction of M and P in terms of productivity and this misconception should be corrected. The idea of trying to work harder is also wrong.

There is absolutely no guarantee that P-level improves without the commitment to consistency with the ST and methods. Those who do not clearly understand this idea think that P-level improves by just trying harder or with mental strength. Therefore, I urge you to reflect on the approach for P-level improvement.

Target P-level for the third intensive promotion is set with the following suppositions:

- Pace possibility is 130%.
- Delay allowance (5%) is zero.
- Fatigue and personal allowance (total 10%) is estimated as a possible minimum 5% for them, for example.

Then, calculate possible reachable P-level as a target for each performance measurement, work shop unit, and so on. The figure will be 120–130%.

Improving performance does not mean increasing working speed/pace. There is a way to evaluate workers' pace with its leveling with four elements; that is called the Leveling methods, or Westinghouse methods, or Lowry, Maynard, Stegemerten (LMS) methods. There are four, but the two of skill and effort are much more effective than the other two of consistency and working conditions. This means worker's working pace depends on them. The definition of skill and effort follow.

As explained, working pace is (skill × effort). Skill means ability to follow standard methods. So, skill level fluctuates very little. Effort is eagerness or an earnest attitude for assigned work. So, it fluctuates very widely all the time. According LMS, range for skill is +0.14–0.22, its range is 37%; effort is +0.13–0.17, its range is 30%. The skill of a worker is steady even if there is a wide range among workers, but effort fluctuates for a worker and the range is about 30%. This range needs to be managed well through the FM's instruction and/or supervision with P-control.

So, there is a necessity to measure, supervise, and instruct workers by the FM; then it is possible to find effective points to improve workers' performance.

7.3.3.3 Reasons to Lower Performance

The losses at shop floors are the responsibility of the workers and the FM and are unavoidable. Management believes workers' responsibility for losses is not the main share of the losses. Actually, however, the workers' share is more than 50% of all losses on shop floors. Typical losses that workers are responsible for are the following:

- Over allocation of workers.
- Standard number of workers is defined by the ST but the FM ignores the standard.
- Disregard of standard methods.

A question is whether workers and the FM know SOPs well or not. SOP is there in the shop floors; sometimes they know it but don't follow it precisely, other times they ignore it completely.

Indirect work that belongs to materials handling staff is done by workers without any question of whose responsibility it is. Synchronized lines should be controlled by conveyor speed and pitch mark, otherwise workers can't recognize whether their working speed is reasonable or not.

Minor idle time. One or two minuets of micro idle time happens on shop floors. Such minor idle times typically happen at opening time, and before and after short breaks and lunch break time. Such idle times also happen at occurrences of changeover.

Ineffective U of machines or facilities. How are machine speeds set; what is the basis of it? SOP includes these standards. Let's take a simple example of handling a water bucket. How much water is an effective handling volume? A heat treatment facility, for example, often misses this point of U losses. The number of works and distance between works in the heat treatment furnaces need to be considered.

Operation pace. The pace or speed of workers should be examined. First of all, workers' working pace should be standardized. Machine tools are defined well in the MHD. The MHD includes recommended cutting conditions, the numbers of operation to repeat for painting, and the number of pieces that should be processed at one time.

These losses are workers' responsibility, but originally they are the FM's responsibility, because they are a result of poor supervision and instructions. Remember, the FM is a key point to reducing these losses.

7.3.3.4 FMs Control Their Shop with Precision

At the beginning of P-control, the FM's variance of performance is wide but the variance becomes very narrow when those performances increase to more than 100%. Standard deviation among FM's performance variances is improved from 12 to 4 in the example shown in Figure 7.15. This change of variance comes from how precisely the FM's control their shops. Exactly the same change can be seen among many measuring workstation units. It is important to point out that FMs must always control their own shop precisely. In the example, standard deviation among the FMs' performance was 12 at beginning of the week then improved to 4. This means the FMs did not control their workers' performance well; the FMs did not take enough care to keep a steady P-level through adequate supervision and instruction at their shops. The FMs' work contents did not adequately follow standard methods and times which had been set as company standards.

We defined control as staying on course, adherence to standards, and prevention of change. A plan requires that we adhere to the plan, which in turn requires

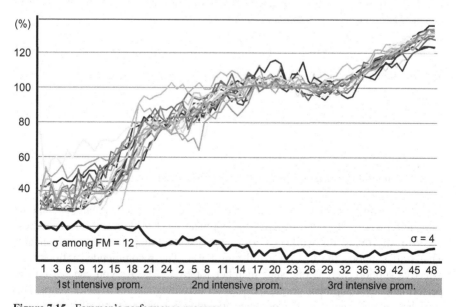

Figure 7.15 Foremen's performance progress

Figure 7.16 A definition of control

> A definition of control –
>
> "staying on course, adhering
> to standards, prevention
> of change."
>
> (Juran 1995)

that we *keep the plan on course*, meet the targets or goals of the plan, and prevent outside forces from damaging the plan. Control is more nearly synonymous with *preventing a change from planned performance* (Juran 1995). Juran defines the meaning of control in Figure 7.16.

7.3.3.5 Mixing of Responsibility of Workers and Foremen

At the beginning of P-control, time consuming contents belong to both workers and FMs. Figure 7.14 illustrates that condition. U due to nonworking or idle time reasons decreases from 90 to 80%. This means for the first 3 months classification of two responsibility categories were mixed. After those months, U did keep a steady level of 80% and only labor performance was improved. This is why total productivity improvement just follows labor performance improvement. This is a simple but very important matter because management believes room for productivity is in U rather than labor performance. U is not a key factor for productivity improvement when a company does not implement work measurement. There is no change of U except starting at 3 months.

7.3.4   Keys to a Successful Performance Improvement

Performance improvement can't be made with attempts to motivate FMs and workers.

7.3.4.1 Special Organized Support Activity as Intensive Promotion

Performance improvement is done by the FM and a specially organized support staff that has a background in industrial engineering. FM does not know the best ways to improving his workers performance, so the support staff is organized. That staff supports the FM on a full-time basis. The recommended ratio for an effective support staff is one staff member for two FMs. As you can see on the performance development graph, the support activities last for 1.0–1.5 years. The staff has a background and experience in industrial engineering of motion and time study. They are familiar with the concept of contents of ST. They can advise on how to follow standard methods and time for any operation. There is a big difference between actual time and ST. Improving the difference between them improves performance itself.

7.3.4.2 Reinforcement of the FM

The FM is absolutely the key person for improving labor performance. It is recommended to strengthen the FM. The FM is expected to supervise and instruct their workers in every way when they are at their shop. The FM must spend at least one third of his shift hours on supervising and instructing his workers before effecting improvement (Figure 7.17). For successful P-control, the FM's activities should be improved as much as possible, such as by spending more than 80% of his time for supervision and instruction, at least in his own shop area. The FM or supervisor is the key person who instructs, supervises, and motivates workers directly.

The following four subjects are key points to improving workers' performance by the FM. The qualifications of a successful FM under these circumstances are:

- to allocate workers based on a reasonable standard number of manning;
- to instruct workers on how to avoid their idle time;
- to instruct workers individually based on standards; and
- to motivate workers in every way.

Kadota Takeji wrote about the image of a good FM as follows:

The FM must stay in the shop at all times to supervise his workers.
The FM must minimize the time spent for incidental activities such as clerical work, meetings, and chasing after delayed parts.
The FM must supervise his workers directly.
The FM should not leave his supervising work to his subordinates, such as group leaders, while spending his time on incidental activities.

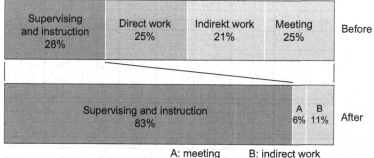

Figure 7.17 Foremen's activities

The FM must supervise and instruct his workers individually and specifically.
Individual and specific instructions are more effective than abstract lists and talks to the whole group.
The FM must be strong.
To develop a capable FM, it is important that the right person be selected and trained well. In addition, the FM must be delegated authority which will give him sufficient influence over his men (Kadota 1968).

7.3.4.3 Useful Performance Report

Table 7.4 is an example of a weekly performance report. Points to be considered in the report are performance (labor performance, U of FM responsibility, and percentage of FM's time spent on instruction at his shop. Labor performance is the main subject of the weekly report; improvement is expected. U is normally 80% for FM responsibility, 90% for management responsibility. This means 20% non-working time occurring in a shift is due to the FM's responsibility and less than 10% is due to management responsibility. Higher or lower than 80% of the FM's U are not acceptable ordinals; lower than 80% might be due to unreported

Table 7.4 Weekly performance report

subjects			plt.Mgr.A	div Mgr. A-a	div. Mgr. A-b	plt.Mgr.B	div. Magr. B-a	div. Magr. B-b	div. Magr. B-c
number of workers			138	69	70	271	58	134	79
performance x utilization	this W.	%	47	50	43	40	41	42	37
	last W.		44	47	39	37	35	38	36
	last M. av.W.		35	39	31	35	34	38	31
performance	this W.	%	63	67	58	57	60	58	52
	target		0	0	0	80	80	80	80
	last week		60	64	54	53	52	56	50
	last month Av.		45	49	41	44	44	46	39
utilization	this W.		74	75	74	71	68	72	72
	last week		74	74	73	69	67	68	72
	last month Av.		78	79	77	81	78	82	80
manager	this W.	%	96	97	95	98	98	97	98
	last W.		96	96	95	96	98	96	95
FM	this W.		77	77	78	73	69	74	74
	last W.		78	79	77	73	70	71	77
input man-hours			5,732	2,794	2,938	10,599	2,134	5,371	3,093
working		MH	4,258	2,082	2,176	7,556	1,453	3,867	2,227
non working			1,381	655	726	2,967	666	1,471	759
top three	1		material	materials	materials	materials	others 1	materials	materials
		MH	697	268	429	1,475	473	863	474
	2		others 1	others 1	others 1	others 1	materilas	others 1	others 1
		MH	322	213	109	972	138	352	147
	3		others 2	training	others 2	machine	others 2	machine	others 2
		MH	99	52	86	119	34	87	46
missing man-hour		MH	92	58	35	113	16	47	50
total standard time			2,687	1,385	1,260	4,292	873	2,242	1,152
standard time		MH	792	503	289	1,270	316	644	310
standard time value for not set			1,895	882	972	3,022	557	1,597	842
FM: % of supv. & instruction		%	96	95	96	96	97	95	97

PERFORMANCE WEEK REPORT	17 w	from 07/10/21 to 07/10/27	date 07/10/29	No. 17

idle time. A possible reason for those percentages could be the FM's missing results of supervising. FM has to be asked to give adequate instruction and supervision; it is also important that the FM always stays in his shop to supervise and instruct his workers. A lower than 90% result has to be managed well by management at the weekly meetings.

7.3.4.4 Report Meeting

Report meetings should be weekly; between 1:00 pm and 2:00 pm is a typical time to meet to discuss the previous week's performance. The division manager chairs this meeting; all FMs and support staff of the division should attend. The meeting agenda is:

- reviewing last week's performance;
- causes of changes;
- this week's performance target; and
- actions to meet the target.

Especially improvement depends on action contents. Actions that the FM is expected to deal with must be specific and intended to change the current actual situation; they should not be a general sort of statement to his workers. Each worker should be able to implement the specific actions on their own. Those actions must indicate exactly what change needs to take place, for example, changing the working area layout. It is appreciated if the FM can estimate what performance improvement percentage can be expected with each action and also how it will contribute to the FM's weekly performance. Small details and actions are not covered in the meeting. Action contents discussed at the meeting are expected to be as concrete as possible and the actions taken or not taken must be reviewed at the next week's meeting. The FM and the manager recognize well that actions are the primary concern, and that performance figures depend on those actions. Planned actions must be not just inspirational such as "to do our best" and so on, but also concrete and specific, e.g., "to modify working area layout to reduce unnecessary movement/steps of workers". The manager/chairman should have prepared specific agendas in order to activate the whole meeting in such a way as to make every attendee have a sense of crisis about improving performance. Having a meeting means getting together and having a discussion.

It is essential for supervisors and managers to understand their mission to instruct every worker.

At shop floors without ST, there are working methods which are totally dependent on the worker's own working methods and performance level. Those workers have no way of knowing about effective standard working methods and standard working paces that are based on world-wide standards. I would like all supervisors and managers to ask yourselves if you are training each worker on the proposed methods in detail, and if you are paying attention to each worker's performance on a daily basis. It seems to me the problem lies there. Therefore, there

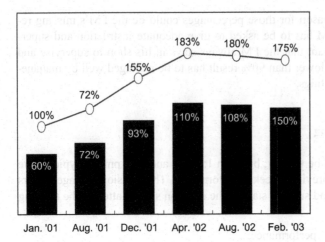

Figure 7.18 Performance improvement at processing machine

is no point to question a fluctuation of 10–20% or the incompletion of TCT without doing what you are supposed to do. What is required now is the fundamental change of performance and the implementation of performance management. If we do not have those fluctuations, the management may not be required on the shop floors. Once again, implementation of new methods can be completed regardless of the current P-level and I think it is important for all the staff in the manufacturing department to always ask a question about your own responsibility in your position.

P-control is not only effective to improve workers' performance, but also machines and/or facilities. For example, a plant of a company is a totally mechanized process-oriented plant. Their performance started at 60% of ST and reached about 110% within 2 years (Figure 7.18). Performance improvement can be identified as increasing production per hours. Operator-tested operating speed of machines/facilities is introduced as ST. To follow ST a lot of changes must be made of past methods which are not accepted or included in ST. There are particular reasons why P-levels as low as 60% happen at the beginning of P-control. If ST was set with simple measurement and setting, 60% of performance was never measured. How high the level of the engineered standard should be set is up to the industrial engineers. A performance level of 60% was never discovered without such a ST.

References

American National Standards Institute – AIIE (1983) Industrial engineering terminology. Wiley-Interscience, Hoboken, NJ
Barnes RF (1980) Motion and time study, design and measurement of work, 7th edn. Wiley, New York
Bayha F, Karger DW (1977) Engineered work measurement. Industrial Press, New York

Juran JM (1995) Managerial breakthrough: The classic book on improving management performance. McGraw-Hill, New York

Kadota T (1968) PAC-Performance analysis and control. J Ind Engng 19:407–411

Mundel ME (1978) Motion and time study improving productivity, 5th edn. Prentice-Hall, Upper Saddle River, NJ

Quick JH, Duncan JH, Malcolm JA Jr (1962) Work factor time standards, measurement of manual and mental work. McGraw Hill Books, New York

Sakamoto S (1983) Practices of work measurement. Japan Management Association, Tokyo, Japan

Society for Advanced Management – SAMUS (1954) A fair day's work. New York University College of Engineering, New York

Chapter 8
White-collar Productivity

This chapter describes how the industrial engineering mindset and tools are good enough to solve productivity issues, but there is no concrete technique utilized for office/white-collar productivity. Managing office productivity (MOP) is an effective technique to apply and receive effective results. How does office work differ from production work? White-collar jobs face tough resistance from employees as managers struggle to be attuned to productivity improvement activity. That resistance is a result on a misunderstanding about productivity that entails mistaking productivity for rationalization.

8.1 Managing Office Productivity: a Tool for White-collar Work

There has been a lot of thought given to the challenges of white-collar productivity over the years. An effective solution has not been found; some respond to the challenges by giving up and/or deciding it's an information technology (IT) issue, as if that were the only solution. The industrial engineering way of thinking and its tools are capable of solving this issue, but there is no concrete technique that is applicable for improving office/white color productivity.

MOP is an effective technique to get effective results. MOP is based on experiences in production areas with industrial engineering. There are three subjects in MOP regarding three dimensions of productivity: M, P, and U. They are called M-MOP, P-MOP, and U-MOP. The outline of M-MOP is introduced below.

Regarding office work productivity, the objective has to be evaluating or reallocating the number of workers instead of reducing time value based improvement. It just becomes nonreal gain. Organization change is required after reducing work contents and allocating a suitable number of workers.

Management cannot know what a reasonable size of human resources in office work should be, but they question it and try to find a technique that will solve it and quell management's concerns. Management has fundamental doubts about increasing the number of office employees and tries to find ways to get relevant information about the situation. On one hand, what happens in offices? IT is integral to office work but not only does it not reduce the number of office employees, it actually increases the number. People who have an interest in IT implementation say increasing productivity is one of the purposes of the implementation; however, it does not in fact reduce the number of employees. The person who has an interest in IT implementation says that its merits are not productivity improvement but rather improving the quality of information. Management is missing an opportunity to increase productivity due to the weakness of management's own vision regarding IT and/or office productivity. Management simply accepts IT investment and just follows the IT vender's recommendations. Not only might IT investment without reasonable examination fail to affect productivity improvement but it also involves a lot of expenditure, not only for the first implementation but also subsequent modifications and maintenance.

The ways that office work differs from production work include:

- Work contents are wide ranging, complex, and not easy to recognize.
- The service level as a work result is wide ranging.
- Standardization of office work is difficult and not important.
- Office employees think the OP of office work is the same as production OP, but in fact they are different.

Experiences gained from the shop floor for increasing productivity, particularly methods engineering, can be effectively applied to office work, as discussed below.

A lack of understanding regarding productivity is common in the office setting. When office workers want to improve productivity they simply increase the speed of accomplishing office tasks, but then they cut their working hours. How to rectify such a situation?

8.2   Feasibility Study for Office Productivity

White-collar areas are subject to very tough resistance from employees when productivity improvement activities are discussed. But that resistance can be a result of a kind of misunderstanding about productivity; which sometimes confuses productivity with rationalization. A top manager who was doing a good job of steering a productivity project in a company, replied to the labor union that "we do not precede revolution but evolution with innovative ideas". Productivity improvement experiences in white-collar areas are very poor, so misunderstandings happen. This book is about the objectives of productivity, work contents, and the contribution of those works to company performance. For solving the challenges regarding productivity and the white-collar employee, the FS has the power to illustrate what productivity improvement means. As a result of

Figure 8.1 MOP-FS result of productivity improvement

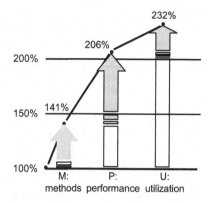

productivity improvement, the time that is necessary to complete tasks and the amount of human resources required are automatically reduced. General improvement principles are the four principles of improvement. The best improvement is the elimination of the purpose (what) of current work. Computerization without work contents improvement is just improvement of "How"; work contents just done by computer.

Collecting data for a FS is gotten with WS. A WS of MOP is done as self-reporting regarding specially designed report contents. The issues regarding reporting contents are BF or AF classification of work, whether the work takes place every month or in a particular month, whether the work was previously or suddenly assigned, and the possibility of computerization. Productivity possibilities are estimated for a case in Figure 8.1.

8.3 Methods of Managing Office Productivity Outline

MDC at shops marked effective reduction of workers without IT (information technology) or small expenditures. MOP is the application of MDC for office/white-collar productivity.

Table 8.1 shows the results after application of M-MOP (Sakamoto 1985). After implementing new working methods, redundancies became apparent and human resources were subsequently adjusted according to management's perceived needs. The BF, percent is normally less than 50% in the office, the rest is AF or waste. Waste is in an office setting is similar to that on a shop floor; one just waits for an example of it, *i.e.*, it can never be *found*. I believe this situation is fairly obvious. This is why thinking about waste elimination is not a useful way of thinking; waste elimination seems like an easy improvement, but in fact is limited at best in its effectiveness at improving office productivity. The design of new working methods should be developed with M-MOP; it creates the kind of effective results that management is interested in.

Table 8.1 M-MOP result of productivity improvement

Company	Allocated number of employee			Productivity improvement (%)
	Before	After	Reduced	
A	45	27	18	167
B	187	131	56	143
C	86	56	30	154
D	163	102	61	160
E	92	48	44	192
F	37	21	16	176

Unique techniques of methods engineering such as MDC for the manufacturing area can be effectively applied to office work. Theory and a fundamental approach do open office productivity. Issues regarding application follow.

- The point of applying MDC to the office setting is that it is necessary to measure processes and activities at the unit level.
- Measuring workload can result in three variation values: maximum, mode/average, and minimum.
- Yearly base measurements of work contents are collected and set as the WU and WC.
- Work contents are divided into two categories depending on whether the work takes place every month or in a particular month.
- Workload is always counted with the allocated number of workers (if the objective is evaluating manning suitability.

Regarding office work productivity, the objective has to be evaluating or reallocating the number of workers instead of reducing time-value-based improvement. It just becomes a nonreal gain.

Organizational change is required after reducing work contents and allocating a suitable number of workers. Workload can be calculated using WU and WC and two points need to be considered. (1) The practical load is decided depending on approval processing time (APT). (2) This is a very important point regarding converting workload to the number of workers. Based on M-MOP experiences, APT is best set at 1 month of working hours. It looks like much work should be processed at a special time, such as until lunch break, immediately after a given event, on special days of the month, and so on. But almost always the theoretical time limit is within 1 month, so *APT is normally set at 1 month.*

A key step is defining the current model. Office work is absolutely not standardized. This means not only management but also everybody in the office doesn't know the standard of their working methods. They just do their jobs according to their own personal working methods.

Setting WU and WC for office work. Workload is calculated by multiplying WU by WC . It is not recommended to standardize work contents the same way it's done on shop floors, such as getting work contents and time values using direct time study (DTS). There is no need to do such a level of standardization for office work. Measuring a unit of office work is bigger than the operation; activity is good enough. A point is self-reporting by all employees in design objective divisions regarding WU (process or activity level base) time value with three values: mean (m) or average, optimistic (o) and pessimistic (p) value. Remember optimistic is not the minimum and pessimistic is not the maximum, they are based on normal conditions; they do not mean an irregular or rare time value. WC also gets three values as well, mean (m) or average, optimistic (o) and pessimistic (p).

$$\text{Representative value for WU or WC} = [p + 2m + o]/4$$

These WU and WC figures are self-reporting but good enough on accuracy and a practical way of thinking. Such a wide range of variation of WU and WC are one of the typical issues in office work. The actual working hours do not fluctuate even with some kind of irregular work with a time value that fluctuates a lot. Employees control their work within normal working hours everyday, as you know.

Transfer work load to number of manning. Workload can translate to a number for allocating workers as follows:

$$\text{Number of manning} = (WU \times WC = \text{work load})/(APT)$$

This is a very important point regarding converting workload to a number of workers. Based on M-MOP experiences, APT should be set at 1 month of working hours normally, but there are very rare exceptions where there is a special time length to process certain work. It looks like much work should be processed at a special time, such as until lunch break, immediately after a given event, on special days of the month, and so on. But almost always the theoretical time limit is within 1 month, so APT is normally set for 1 month.

Let's introduce an example to understand the meaning of APT. It is assumed that 100 copies of a 50-page document are needed. It takes 120 min to make that many copies, and 30 min are allowed to complete all the copying. Since 30 min is the APT, and 120/30 = 4, that means 4 workers are necessary to do the copy work and meet the 30-min APT. Even with the same amount of workload, the necessary number of manning changes.

How to handle occurrence of operations that depend on a particular week or month. There are two kinds of work, which are defined by when they occur: (1) on a weekly or nonweekly basis and (2) irregularly, week to week. It seems like these kinds of operations happen frequently, but in fact it is less than 20–30% of total work, a fact might escape the notice of office employees. This condition means that only every-month work should be subject to methods design; it means a week of improvement can apply for all weeks in a year. It is enough for methods design to use a minimum in part of design activity. There is no need to be concerned with irregular occurrences; every-week or every-month work is good enough for designing new methods. Nothing is done about any other kinds of work, even a change of methods, because those kinds of work comprise a small percentage of the workload. If unusual things happen because of these other kinds of work, the workers are capable of handling it themselves.

The performance dimension is not a realistic objective area of productivity. The performance dimension is not a realistic objective area of productivity in the office due to the difficulty of determining work measurement standards.

Utilization for the office. Plan work of a day with operations or activities and occurrences based on WU and WC. Office employees know themselves that working hours and work contents are the basis for quantification. An action for this matter is balancing work load and working hours every day. It is accepted that the employee should set the expected P for WU time value. Irregular or unplanned work happens; it is accepted that this also should be dealt with based on self decision. This seems to be a rough approach, but it actually causes no problems in practice and is good enough to find a reasonable balance between workloads not predicted to result in saving time.

Reference

Sakamoto S (1985) Methods of managing office productivity outline. Japan Management Association, Tokyo, Japan

Part IV
Monitoring Productivity

Chapter 9
MBM: Measurement/Monitoring-based Management

Chapter 9 establishes that management should understand that the expenditure of time and money will optimize the effects of productivity. Gold mines of productivity can never be discovered in one dig, but daily measurement promises the ultimate results. Three productivity concepts such as M, P, and U are measured with ST. There is a practical solution for office productivity that can be measured by preparing two categories of measure. One is productivity in behavior/processed workload; another is productivity in purpose/contribution of work.

Management behavior can be changed by management with the help of experiences and hunches that F.W. Taylor wrote about in scientific management about measurement based management (MBM). ST is a necessary matter for the practice of MBM and ST acts as a magnifier regarding solutions. A lot of issues regarding productivity are missed by management; management does not have a magnifier such as ST. Setting ST, training for MTM practice, developing STD, and measuring system development takes time. Management should understand that those expenditures of time and money will have a great affect on productivity. Gold mines of productivity don't appear out of nowhere; daily measurement promises the result.

Measuring productivity requires a high level of expertise with industrial engineering; however, any management from top to bottom must simply insist on the necessity of productivity improvement for competitive advancement, reasonable profitability, and/or profit.

An example of monitoring productivity is introduced as monthly and quarterly productivity reports for production and two measurements for office.

9.1 Monthly Productivity Reports

Figure 9.1 is an example of a monthly productivity measurement report measured by ST.

Monthly productivity report

December, 2009

	measuring items			Manufacturing division			Plant A			Plant B		
				this month	previous	last year	this month	previous	last year	this month	previous	last year
1 productivity	OPM		%	98	95	95	100	97	94	96	94	95
	total reduced man-hours		MH	1.548	-1.298	-639	728	-185	271	820	-1113	-910
		reduced	%	3	-3	-1	4	-2	1	3	-4	-2
	reduced by M & P		MH	1.396	-1.626	166	458	-739	111	938	-888	55
	TPM 1		%	37	35	35	36	34	34	37	35	35
		employee number	no.	644	644	806	248	248	257	396	396	549
2 analysis of productivity	M	reduced man hours	MH	-71	48	107	0	55	29	-71	-7	79
		methods eng.		0	-7	3	0	0	0	0	-7	3
		work simplification		-71	55	104	0	55	29	-71	0	75
		reduced man hours	MH	1.476	-1.674	96	458	-794	89	1.009	-881	7
	P	total performance	%	99	94	97	96	93	93	101	95	99
		operation performance	%	117	119	118	116	121	119	118	119	118
		utilization		85	79	82	83	77	79	86	80	84
		idle items	MH	7.908	9.456	10.023	3.065	3.617	3.948	4.843	5.838	6.076
		1st content		materials short.	materials short.	materials short.	materials short.	materials short.	materials short.	materials short.	materials short.	materials short.
	U	reduced man hours for bench mark	MH	107	361	-815	218	587	172	-111	-225	-987
		difference		-107	-295	888	-217	-521	-100	110	227	989
		set-up hours	MH	-107	-295	888	-217	-521	-100	110	227	989
		difference, application ST		0	0	0	0	0	0	0	0	0
3 production in ST	total earned hours as standard time		MH	45.167	44.569	49.056	16.457	15.406	17.227	28.710	29.163	31.830
	effective work contents			42.368	44.880	48.409	13.790	15.960	17.239	28.579	28.919	31.170
		products		34.315	33.716	37.245	13.790	13.093	14.372	20.526	20.622	22.873
		others (setup, others)		10.829	10.869	11.490	2.667	2.346	2.703	8.163	8.524	8.787
	ineffective work contents			0	0	0	0	0	0	0	0	0
	MH/one operator/day		MH/old	7,5	7,0	7,5	7,5	6,9	7,3	7,4	7,1	7,7
	input/consumed man hours		MH	45.519	47.199	51.147	17.069	16.469	18.441	28.450	30.730	32.706
		overtime per month, worker	H	19	18	10	27	28	15	14	12	6
		number of worker under P. cont.	no.	325	327	344	105	107	114	220	220	230
4 labor cost	total paid wage		JPY1,000	140.237	142.652	152.499	50.726	53.870	54.166	89.511	88.782	98.282
	actual wage rate		JPY/MH	3.024	3.068	3.019	3.275	3.271	3.176	2.880	2.952	2.929
	reduced man-hours as labor cost		JPY1,000	4.747	-3.897	-1.678	2.384	-565	853	2.362	-3.332	-2.532
	ULC, unit labor cost		JPY/MH	2.581	2.571	2.698	2.816	2.711	2.803	2.447	2.491	2.638

Figure 9.1 Monthly productivity report

The shop floor level of productivity was improved; however, the affect on the whole plant including indirect areas is very limited. Two productivity measurements are prepared on a monthly productivity report. These are OPM, operational productivity measure, and total productivity measure (TPM). OPM is shop floor productivity, TPM is productivity measure covering indirect areas who support shop floor productivity. Figure 9.1 shows a large gap between them. A vital solution is to narrow the gap between them to improve office productivity.

There is an interesting result on the monthly productivity report. It is "work simplification" in the column for the M dimension. The company is eager to steer work simplification although those results are not nearly enough. A lot of improvement or changes are done but those are too small for changing ST. Generally speaking, those improvements or changes do not have a large affect on productivity. Those activities do affect a few other matters regarding corporate performance.

There can be no objective judgment about productivity, and whether methods have been improved or not, if it is not measured with ST.

The way three productivity contents, M, P, and U are measured with ST is discussed below.

M, methods dimension: accumulated difference of ST between before and after about changing methods. It also is divided into two parts: management and staff engineers and work simplification. For example, the standard before Improvement is 12 man-hours and 6 man-hours. Improvement methods are calculated as $12/6 \times 100 = 200\%$, double of productivity change of methods improvement. Reduction of the ST for a particular operation is the effect of M, methods change. Management and staff engineers is methods change with management and staff engineers contribution, such as industrial engineer, production engineer, quality, and so on. Work simplification means an improvement effect with the participation of management such as a suggestion plan for a small group activity SGA. A significant part of measuring M effects are only measured as reduced man-hours of the ST base difference between before and after the change. The difference accumulates based on average count of production value.

P is ST divided by actual working hours. Such as 5.00 MH/7.00 MH, man-hour = 71%. A 100% performance level meant a worker just followed standard methods precisely with standard pace. Nobody evaluates or measures workers' performance without engineering ST.

There are three measurements of P; they are total performance, labor performance, and U and are included in the weekly performance report.

Total performance

$$= \text{ labor P} \times \text{U}$$
$$= (\textstyle\sum \text{produced products, parts in ST})/(\textstyle\sum \text{consumed labor man-hours})$$

And the percent of change gains (+) or loss (−) in man-hours are calculated.

The top three categorized reasons of down/unproductive/idle time in man-hours are introduced. Management and/or support staff can know the conditions of P.

Table 9.1 Application standards: ABC standards

Standardtime	Production per hour (pc/h)	Cycle time (min)	Line balancing (%)	Standard manning (number)
A	60; 100%	1.0	95	10
B	75; 130%	0.8	90	14
C	43; 70%	1.4	97	8

The last is U. U contribution to productivity, which is not possible to measure simply like M and P. It is not easy to measure this dimension of productivity effect, but it is possible to measure the effect of U function. It measures effectiveness of production and planning is expected to reduce set up/changeover. However, it cannot measure absolute values like M and P functions. How much effect for productivity is made with production planning? The number of production opportunities and its total set-up hours in a month is productivity losses; stoppage hours is the wrong effect for productivity, quality loss of production as well.

So, practically thinking is as follows. Reduce set up/changeover opportunity as little as possible. A zero set up/changeover is impossible; this is why the practical minimum opportunity within the past-6-months record is set as a benchmark. The difference of man-hours between a month and the benchmark is U effect in productivity.

Another issue reducing set up/changeover man-hours is possible according to the methods change of set up/changeover such as SMED, setting, which is measured as M dimension because ST for set up/changeover is changed.

Workers are asked to do an operation that does not need ST. The application standard (or A/B/C standard) or temporary use standard are set in order to escape the minus effects of the condition such as production speed, quality control, preventive maintenance, and so on (Table 9.1) (Sakamoto 1983).

The application ST sets for different production speeds per hour, for example, without inefficient work (lowering performance). Otherwise, the results of measured performance declined in order to change production volume or speed. Such a reason for declining performance is not a reason to know declining performance. Application ST in measuring productivity escapes such a reason for performance but this reason is due to the production planning and control department. Similarly, there are opportunities that have a worse effect on performance due to support staff responsibility, for example, a work station put on original work contents due to checking quality for temporary provisions of quality defects results in increasing work stations, increasing manning. Similar things happen due to facility maintenance, such as a machine has not broken down but production plans demand a continuation of production with adding an operator for keeping production volume of the machine condition.

These kinds of changes should always reflect the ST as the application standard; temporary standard in man-hours is the part of the U dimension effect on productivity. Why is ST effective? There is no way to manage productivity without measurement by ST that is based on world-wide common standards such as MTM. Methods, whether good or bad, cannot measure without objective standards. There is a traditional way to evaluate methods effectiveness by comparing actual times. That is, comparing the difference between two methods by measuring the time of those two practical operations. But it is dangerous because those actual time results include and accept those workers' performance results. The effectiveness of an operation between the current and improved situations is only possible to measure without the workers' performance effect. That measure is ST.

9.2 Two Measures of Office Productivity

Misunderstanding productivity is a common condition in the office setting. Office workers understand productivity as meaning increasing the speed of completing work and then cutting working hours.

How to solve such a current conservative condition for productivity?

There is a practical solution for measuring office productivity that avoids office employees' lack of understanding. It can be managed by preparing two categories of measurement (see Figure 9.2).

One is productivity in behavior/processed (measure A) workload; another is productivity in purpose/contribution of work (measure B). The first one is the number of produced products divided by the number of consumed man-hours, for example, and measurement of productivity in processed workload is not only the number of produced products but also the quality of the product. So, the measure A is just measuring one part of productivity. Productivity with the measure of A is a main measure of manufacturing area productivity. It is also important in the office, but productivity measure Y is much more significant for office productivity. Let's take a simple example at an engineering department. The number of

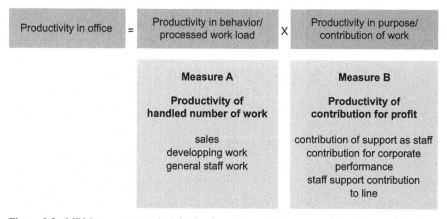

Figure 9.2 MBM: measurement/monitoring-based management

completed drawings by an engineer is the productivity of the measure A. The engineer makes more drawings than others; is it comfortable for the department? The answer is no, because if the engineer's drawings include a few defects in the production stage, for example, is he a good engineer from a productivity point of view? The answer is no. This productivity is the measure of productivity in purpose (B). So, an engineer took time to complete assigned drawings, but his reputation in the manufacturing shops is that too few of his drawings have no problems. The conclusion for white-collar area workers' productivity is A times B. At the beginning of challenging white-collar productivity, these two productivity definitions must be precisely with those people and industrial engineers. Remember measuring productivity is a kind of professional area.

Reference

Sakamoto S (1983) Practices of work measurement. Japan Management Association, Tokyo, Japan

Part V
Keys to Success for Improved Management

Chapter 10
Changing for Productivity

Chapter 10 describes how active managers are interested in development factors and passive ones aren't. There is a saying that some people cannot see the forest for the trees. Correct understanding of methods and performance is required. Developing an innovation-minded view with regard to performance control is a significant matter for successful productivity improvement. An innovation-minded view means being able to take approaches that consider what "should be" or "can be" done from the standpoint of each position. There is an old four-word saying, "Zui-Sho-E-Shu" in Japan, which means you should take the initiative no matter where you are.

10.1   Creation of New Methods in MDC

The following are practical points for management to consider when trying to create more effective management regarding productivity. They are a result of experiences with MDC and performance control practices.

Management's mission is to change the process from "given condition" to that condition which will supply the required results. So, productivity improvement through industrial engineering activities are "opportunity profit". Whether or not to utilize the profit as corporate profit is completely up to management (see Figure 10.1). The profit can be made larger than opportunity profit or smaller depending on management.

10.1.1   Manning Number Depends on Production Rate

The assigned work is calculated with work contents and time and divided by production rate TCT to find the required manning number. In other words, even with

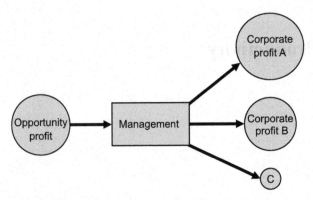

Figure 10.1   Corporate profit is dependent on management

the same workload, the required manning changes almost proportionally to the TCT. Changing of the TCT depends on product inventory based on sales results, rather than on the needs of manufacturing. Therefore, for production planning, production volume per day or hour, that is TCT, is based on expected sales and inventory. The production volume automatically determines the required manning. It is a matter of course. How have you been responding to changes in required manning following changed TCT? The number of workers might not be employed workers, whether permanent or temporary, depending on TCT changing. Instead of such an adjustment, the FM and/or manager might manage to solve the problem "tactfully" by finding workers somehow from the workforce of your own or other departments.

The appropriateness of the new manning according to a certain TCT is evaluated at the time of the MDC proposal. The required manning can be calculated from the TCT at the time of implementation of the new work method. For example, assume that an improvement method is proposed to reduce ten workers to five workers at TCTm, which is a manning reduction of five workers. But TCTm has changed to TCTn at the time of implementation, for example. Before the MDC proposal, TCTn required fourteen workers instead of ten workers (appropriateness not verified). So a production manager says it is impossible to reduce five workers, and moreover, it is a lack of four workers (14 − 10 = 4). Therefore, the reduction of five workers is only feasible in the condition of TCTa production management. What do you think of this comment? Is this right from the management point of view?

The answer is "No". The proposed reduction should be implemented immediately because: (1) It is not clear how the lack of four workers was compensated at TCTn before the MDC proposal, and (2) If a 50% reduction is possible by the MDC proposal, fourteen workers before the proposal can be reduced to seven workers with the new method, and only two additional workers, not four, are required at TCTn. Again, it is a matter of course.

10.1.2 "What You Can Do" vs. "What You Should Do"

In the Manufacturing Department, a variety of different problems which are hard to predict occur on a daily basis, and the department is expected to respond immediately. Therefore, it is natural that middle management is staffed. But response to these problems is not enough. It is nothing special or beneficial to corporate performance. I classify this as a "maintenance factor" and a new contributor into the "development factor". Active managers are interested in development factors, and passive ones aren't. But I don't say that either of them does what they should do as a company. That makes it even more complicated. A manager who is interested in maintenance factors holds development factors inside, sticks to conservative approaches, and delays realization of achievements. There is a saying that "some people cannot see the forest for the trees". There are some things to be done by priority in a company, regardless of the factory's actual conditions or the maintenance factors' situations. In MDC, a criterion for completion, which is different from what the company should do, is needed depending on the difficulty in the practical phase. These are the suspicious points in our MDC activities. "Is it good to judge giving maintenance factors priority?"

If the contents of MDC proposals are exhaustively specific, they are easy for managers to practice. But in reality, it is impossible to prepare them because a company divides labor to aim for all-round achievement. So, it is important that everybody raises their level of communication and collaboration higher than usual, not dividing planners and practitioners of new MDC methods, and tries to improve the performance of workers. Management or supervisors are expected to have more interest in "what you should do" than "what you can do".

10.1.3 A New Standard vs. New Practice

The new standard proposed a new work method that results in higher line balancing, for example, as high as 100% and a balance loss of 5% or less. But these are calculated results based on a standardized workload (WC and WU). The standard is imagined with nearly an equal workload, that is, "tact time" among work stations, where completion of one unit almost coincides with completion of another unit. On the actual site, however, there is a pile of WIP among workstations, which is quite similar to before the improvement except for reduced manning.

One problem is the working time that is measured by stopwatch time study for the model setting in MDC. In short, the workers' performance was low and fluctuated widely. If the performance was high, the workers would have been working with acceptable high performance, minimizing fluctuations. This is because performance standards are limited. The SLB and DLB difference is a good example of divergence between a new standard and a new practice. Managers and supervisors are expected to care about this on a daily basis.

10.1.4 MDC Practice Is Not an Objective

An objective to doing MDC is profitability improvement.

For the purpose of improving corporate performance and profitability through MDC activities, manufacturing has started to implement MDC projects. First of all, a question must be asked of the managers. For what purpose are you working on the MDC projects to begin with? I deeply regret that some managers cannot answer this question clearly. This problem indicates their lack of understanding of the MDC procedures and their concerns about various issues at the implementation stage.

Here is an example of a very poor understanding of the MDC procedures: When setting a current model for MDC, a model task is determined by 25% (1/4) selection of measurement results, which makes the assignment approximately 20% severer than the actual condition (result). Furthermore, when setting an improvement model, a maximum task of 1.2 persons is allowed per worker, that is, 20% extra work is assigned to a worker. As a result, the proposed manning becomes very severe and difficult to implement.

First, let me explain the technical aspect of MDC. The most important point is that an MDC proposal sets out a "new standard method". It is an "ideal method" (standardized model) considered desirable through the use of IE technology. What is the point of comparing it with the current situation? In other words, the current work condition is not the standard, but just the "reality" based on each worker's own standard, which, we must admit, is quite different from the generally accepted work standard. Standardization of work and workload is essential to achieve higher productivity. If the proposed improvement model work contents and their time seem severe, the current work standard is just too low, and there is no other problem. In successful MDC implementation cases, usually the managers leading MDC implementation at manufacturing sites have no such questions. Managers with such doubts should reflect on their incomprehension of what is going on at the site and their endless empty discussions of figures on paper. Managers asking questions from a conservative and negative view can never free themselves from the current situation.

10.1.5 The Importance of Performance Control: Practical Hints

Correct understanding of methods and performance is required. Methods are stability that is not affected by workers' performance such as skill and effort. Methods defined as working methods are quite cool or self-stabilizing. Performance is measured as workers' time consumed for the methods. However, management misunderstands that those time variations depend on workers' performance. Management has to understand that, but many don't. Rather, they are against new methods or insist they are difficult to follow because they cannot understand the difference of methods and performance. There are no problems with new designed methods if nobody meets operation time of the designed methods. This is why performance control is recommended to complete new designed methods. New designed methods concerning operators' consumed time is a purpose of having performance control, but control of workers' performance is an important requirement for the success of new designed methods.

10.2 Developing an Innovation-Minded View of Organization with Performance Control

An innovation-minded view happens in FMs and workers through performance control.

What happened on the shop floors? The innovation-minded view has happened. To stimulate an innovation-minded view, performance improvement activities supported by industrial engineers are developed by a step-by-step approach. What are the keys to improving performance? It is nothing but that the promotion of an innovation-minded view has been implemented which is to management part of P-control before it started. Having an innovation-minded view means being able to take approaches which consider the responsibility that should be or can be taken from the standpoint of each position. The FMs and workers on the front-line shop floor give thought to what responsibility should be taken from their standpoint and have received the support from support staff members in order to fulfill their duties because they cannot do it by themselves. For example, the workers' work paces did not improve as P-level improved. To be more exact, taking operation of ten products, for instance, before performance control, their work paces were fast for only one or two products but slow for the rest. However, their work paces were improved and they can constantly work at a high pace at present. In order to actualize it, many different actions have been steadily implemented in the Intensive Promotion.

Referring to other cases, the distinction of operation and idle time was not clear at the early stage of P-control, and the workers had a wrong conception that even the idle time was conceived as operation. The distinction of operation and set-up time was also unclear. There are actually numbers of other cases but for each one, the workers have spent their working hours considering their responsibility that should be taken. The degree of the efforts for an innovation-minded view shows in the P-level and the improved P-level is the result of successful promotion of an innovation-minded view implemented by FMs and the workers.

The initial stage of P-control of less than 50% of P-level improved to around 130%. What really happened at the shop floors during the course of the activities?

The answer is simple. It is nothing but the promotion of the innovation-minded view which has been implemented.

Keep in mind that this managers' change are directly proportional to the P-level improvement. There is a need of a company-wide promotion of the innovation-minded view by management and FMs, not to mention for the workers.

There is an old four-word saying "Zui-Sho-E-Shu" in Japan, which means you should take the initiative in wherever you are. Once in a certain company, I saw some part-time women writing addresses on envelopes to their clients. Perhaps they were paid by the number of envelopes they wrote. Their attitudes made me think of something. They were obviously not motivated, and their handwriting looked very poor and was lacking in sincerity. So I said to them, "Look at the company name and position of each addressee. Do you intend your writing to be read by them?" You should not think your job dull and boring. It depends not on what you do, but on what you think. Every one of you should "take the initiative and play the lead" in whatever you do and wherever you are.

Real or Nonreal Gain. Some people cannot see the forest for the trees. There are different ways to see a certain reality, and I am not saying which is better or not. It is important to be aware of which viewpoint you choose. People often see the results of productivity improvement activities from a wrong point of view. It may be right for some aspects or parts of the results. But you should consider whether it is preferable for management activities. It does not matter whether a shortsighted microscopic viewpoint of looking at trees is right or wrong. In management activities, the forest is more important than individual trees. Management should give more attention to the desirability of your viewpoint than to its appropriateness.

Improvement activities should be evaluated not only for their individual effectiveness at production sites, but also for their consistency with the targets established by the company or department. If not, managers are just playing improvement "games". However, such improvement activities with uncertain contributions to corporate performance are easy to find. You should keep in mind that you can never gain a competitive advantage in our industry if you are just satisfied with vigorous improvement activities. They are called "nonreal gain". Management activities need real gain rather than nonreal gain. Can you find any nonreal gain around you? I hope you will always try to learn how to detect nonreal gain and figure out a way of changing it into real gain.

Management initiatives are a key. That is, you still cling to the "old habits" of your conventional improvement approaches. "The present method itself is not a problem. You can freely design a new method for enhancing productivity without any restriction." First of all, you should change your dependence on improvement items to the pursuit of new ideas. The MDC procedure is to create a method of logical workload reduction required in the designing stage. Through MDC practices, I expect the managers themselves to take leadership positions and experience the effectiveness of new approaches for improvement. Implementation without the managers' enthusiastic involvement can yield only minimal required of results. MDC practices are expected to induce changes in the climate, approaches to find solutions, and role sharing among the managers, supervisors and persons in charge. You should aim for significant achievements you have never experienced before and practices of working methods effective in every aspect. A level of just being earnest and hard working is not enough.

10.3   Designing Systems for Success

Companies that have achieved improvements in productivity using MDC have done so by engaging in the following four areas:

- Top down activities
- Design approach
- Full time project teams
- Steering organization

With the fact that it is ensured by the global standards, not by a mere target, there is no doubt we can achieve the target level with steady handling of the problems. In a word, the managers have left an awful lot of room for productivity improvement. "One practice is better than hundreds of disputes."

Improvement results are up to the following formula.

Improvement result/effect = [(adopted/installed technique)
$$\times \text{(IP human resources)}]^{\text{(passion of management)}}$$

The first element in the formula to be determined is the technique to use for productivity improvement. You cannot find any gold when you're at a coal mine. Your attitude must be purpose oriented rather than technique oriented. The second element in the formula is the scale of human resources. The cases introduced here were organized to have 20–50 engineers on a full-time basis. There is a required reasonable size of engine power for a reasonable target of productivity improvement. Small car engines never can move jumbo jet airplanes. The same can be said for productivity improvement. The last element is the passion or enthusiasm of management, especially top management. Conservative or weak decision-making and the attitude of top management should be warned.

10.3.1   Top-down Activities

This refers to making top management responsible for implementing measures to raise productivity. It entails establishing a steering committee for raising productivity within the company, thereby creating a structure that can deliver strong leadership in setting productivity improvement targets, studying proposals, and providing support at the implementation level.

The steering committee includes a member of top management as a chairman, middle management, and staff engineers who are responsible for the project activ-

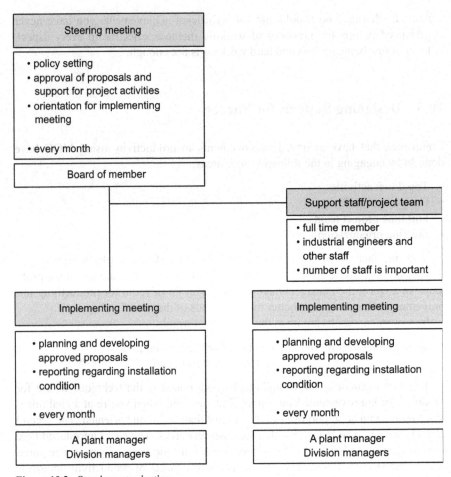

Figure 10.2 Steering organization

ity regarding application and developing industrial engineering techniques for required subjects determined by top management (see Figure 10.2). Management consultants also work well for orienting the steering meeting. Under the umbrella of the steering meeting, an implementation meeting is organized for implementing practice of the proposed subjects by staff engineers. Both meetings are held every month. These two meetings are in a vertical relation, that is, the implementation meeting follows as a result of steering meeting decisions. Top management probably are not good at understanding the details of proposed subjects. However that's all right; middle management tends to have many detailed questions, particularly if they tend to have a conservative attitude. Decisions made by top management need not be concerned with details. This is a simple but important issue with regard to implementation of new methods, for example.

The steering committee establishes the project targets in operational terms. The committee must also establish time, resource, and cost plans as well as quality requirements within the framework of the assignment. The steering committee appoints project managers, project administrators, project members, and a reference group. The project manager is often the "reporter" in the steering meeting.

Against the background of the tasks of the steering meeting, the steering committee should be comprised of people with decision-making abilities. But these people must also have a "knowledge of human nature", *i.e.*, through their own actions demonstrate that they have the answers to the following questions:

- How can participation be created?
- How can high performance be stimulated?
- How can one reward without creating jealousy?
- How should one delegate so that the person feels entrusted but not controlled or exposed?
- How can a tendency to change be created? How can one get people to feel secure in a changing world rather than a static one?

The first improvement proposals from MDC project team are presented at the steering meeting. Then, the implementation meeting prepares implementation plans for the proposals. Compared to the level of its expected result, each improvement is not so difficult in its implementation stage, and its estimated cost is rather low for a labor-saving activity. But, considering the large gaps between the current situations and the proposed improvements, that is, improvement effects, there may be many things to discuss at the preparation or implementation stage. I mean the improvements will take hard work to accomplish. It is only natural. You have not been feeling any inconvenience in the conventional working methods. From the top to the front-line management, everyone has been practicing them as appropriate methods or plans. You may naturally feel an objection to any proposed improvement, especially when a significant result is expected. This is where a mission exists for the management. Their proper activities to handle and manage different problems are anticipated. If such problems never arise, management may not be required. They are expected to think up ideas, devise means, motivate persons involved, and coordinate. Besides, further management activities should be required to contribute to corporate performance and achieve profitability improvement. Not passive and conservative, but active and positive management is the key to success in implementing the proposals.

Management must do well in providing IP for a given condition and achieving an OP that has the required results. If a change in the condition can change the result automatically, management is not needed anymore. If the given condition can be changed to gain the required result, management is simple. That is why management processes are called the "black box" to convert IP to OP.

10.3.2 Design Approach

MDC aims to define the IP state before production work begins, and the OP state after production work has finished, and to develop production methods that link the two. It is not based on current production methods. The time taken by production on the factory floor can be broadly divided into working time and nonworking time. While it may seem obvious, the object of productivity improvement measures is working time. In MDC, working time is divided into BFs and AFs. BFs are those functions that can lead directly to increases in OP.

Although AFs are definitely not waste, their role is to support BFs. Working methods with a high proportion of BFs are taken as being highly productive. In MDC, working method design targets are set for designing new BFs to formulate new ideas. As shown previously, the effectiveness of this approach has led to major improvements in profitability. In my experience, the proportion of BFs in working time in manufacturing units is 40–50%, and in designing new working methods we formulate new ideas and devise specific methods to raise the BF ratio to 70% or more. MDC is not productivity management itself; it should be steering by management with industrial engineers (see Table 10.1).

There is a misunderstanding about industrial engineers and management leading activities or employees participating in activities such as Kaizen and/or waste elimination. Industrial engineers/management and employee participation are not opposing matters. They can be categorized into two parts: (1) motivation-based activities such as waste elimination with employees, and (2) theoretical engineering-based activities such as industrial engineering issues regarding MDC and performance control with engineered standard time. Both subject activities are required. Those activity contents, targets, have an affect on company performance, required expenditures, and change from present conditions are quite different. But remember, fundamental change for productivity improvement which directly contributes to company performance is possible with management/industrial engineers activity, it can't be done with participative management.

Table 10.1 Management levels and effective management

Management levels and target		Effective management
I – low	Better than present	Scientific management or autonomous
II – normal	Compare to outside normal	Scientific management
III – high	Better than outside and higher original	Scientific management and autonomous

10.3.3   Full-time Project Teams

A dedicated team in charge of MDC is vital for ensuring the successful application of MDC for enhancing productivity. Industrial engineers who understand the technical details of MDC need to be employed on a full-time basis. Ideas are essential for designing new working methods with high productivity and the needed ideas are not of the sort those already doing the design can come up with. Rather, they must be new ideas arrived at through BS. Because the designers of new working methods are expected to work with a constant awareness of their targets, employing them fulltime is preferable.

Activities to raise productivity that can make direct contributions to corporate earnings should be implemented; *i.e.*, management needs to take a strong interest in working to raise productivity in a profit-oriented manner. Although this is a generalized comment, among examples of *kaizen* (incremental improvements) in manufacturing units, the majority are so-called empty gains, which do not stand out in business results. Even when the results of the improvements do appear on the company's balance sheet, their contributions to earnings are miniscule. For example, only part of the calculations (or forecasts) for reducing required man-hours or shortening cycle times through work simplification can be spotted in a company's earnings. In contrast, reductions in actual headcounts according to the number of workers needed enables those workers who have been made redundant to be deployed elsewhere, in new jobs, thus increasing the possibility that the changes will show up as profits in the company's results.

What is effective in changing intentions in this way is not grasping after ways to change methods, starting with working practices on the shop floor, but rather a design approach focusing on searches for creative ideas that set and achieve theoretical design targets and have a direct impact on earnings.

Further, if we consider the growth stages of a company, by becoming a highly profitable company through productivity improvements or promoting a shift in production strategy toward technology and business competitiveness, thereby ending dependence on price and cost competitiveness, I believe that the interests of company management will change from growth to maturity. Companies should move away from growth-centered management strategies of expanding market scale and raising market share, and instead aim to become companies that want to exist to benefit society, the community, and its employees and, as such, companies that are respected from outside the company. As a final word, I want to emphasize that the source of the management resources to achieve this, is the tireless pursuit of results-oriented efforts to raise productivity and profitability.

10.3.4   A Key Person Is the Project Leader

Let's introduce more about project organization. A capable project manager is the most important issue for organizing a project team. The success of a project can depend on who is appointed as project manager. More than 60% of the success or failure of project results depends on the project leader. Accordingly, the steering committee's choice of project manager is a critical decision. The three most important skills required of a project leader are:

- collaboration;
- leadership; and
- in-depth expertise in the project area.

The project leader has two main areas of work:

- external, *i.e.*, the relationships of the project with the surrounding world; and
- internal, *i.e.*, within the project organization, and primarily in the project team.

Being outside the normal hierarchy in this respect gives "freedom" to create resources and means of control for the work. But at the same time, this "alienation" can lead to opposition from those on the line who may consider that the project manager "disrupts order". The project manager, however, is an exception. The project manager is both a specialist and responsible for the success or failure of the project. The project manager, using the resources and knowledge of the project team, is to find the best solution and the best path to this solution. The project manager must not become so "blind" as to forget about objectivity.

Why does resistance to change arise? When employees get to hear of the future changes that affect their behavior, problems arise in three dimensions: thought, feeling, and action. The manager must act so that the individuals affected develop:

- understanding instead of doubt (thought);
- trust instead of suspicion (feeling); and
- courage instead of anxiety (action).

The manager must first show the employees respect for their professional competence. The manager must then acknowledge the employees as individuals by being honest, straight, and clear, person to person. Last, but not least, comes participation.

There is a question of who is to be a member. There are a host of factors to take into account. The participants must of course have expertise in the area in question. They must have the ability to analyze and great creative capacity.

In the majority of organizations there are specialists who are important for goal fulfillment but who do not have responsibility for line activities.

Always try to appoint project team members who can work fulltime on the project since it is very difficult for someone to divide their time between the project and normal work in the base organization. There is a risk of conflicting loyalties. Performance will, without doubt, suffer.

Also, someone working fulltime cannot blame their divided work situation. It is easier to control the time schedule of the project. The duration of projects is becoming ever shorter, which has several positive effects.

The tasks of a project manager primarily include:

- operational planning and budgeting;
- professional management of the solution of the assignment ("act as supervisor");
- control, follow-up, evaluation, and reporting;
- collaboration with collateral entities and reference groups;
- information to the base organization; and
- ensuring the project team works effectively internally and externally.

Appendix

A.1 Sequential Activity and Methods Analysis (SAM)

Permission for publishing by The Nordic MTM Association. SAM, Sequential Activity and Methods Analysis, was developed by the Swedish MTM Association in 1983 and is today an official IMD system. It is built on a new way of thinking, mainly:

- sequential purpose-based analysis, increasing the speed of application and making it easier to make, read and understand the analysis;
- minimizing applicator deviations as those cause loss of confidence in the application;
- use of MTM-1 criteria for the choice of type and variables in the system in order to simplify the use of SAM and to eliminate the need of MTM-1 knowledge for applicators of SAM; and
- building the system on a well-defined and scrutinized back-up data, SAM is based on the same back-up as MTM-2.

The Nordic MTM Association appreciates the context in which Mr. Shigeyasu Sakamoto now is publishing SAM. (Note! SAM may not be used without formal training and examination.)

A.1.1 Introduction to the SAM System

The objective of the SAM system is to enable its users to:

- design work methods for high total productivity;
- document work methods in such a way that they can be reproduced with the planned result at any time;
- establish norm times based on documented work methods.

A norm time is the time it will take to carry out a manual task using the documented method at the SAM system norm performance level.

The time unit in the SAM system is called factor.

- 1 h = 20.000 factors
- 1 factor = 5 TMU
- 1 s = about 5.6 factors
- 1 min = about 333 factors

The SAM system's norm performance level is the performance level most people are working at when carrying out manual tasks. When a performance incentive system is used, the SAM system norm performance level usually exceeds 10–20%. Manual work consists of the movement of objects with the hands, in a planned procedure, to accomplish tasks with useful functions. Manual movement of objects follows a consistent pattern of activity sequence: get an object and *put* the object into a planned final position.

The SAM system is based on this activity sequence, which includes three types of activities.

Type	Activity
Basic activities	Get and put
Supplementary activities	Apply force, step, bend
Repetitive activities	Screw, crank, to and from, hammer, read, note, press button

The norm time for an activity varies with the method used. An activity may therefore have one or more variables. For example, the variables for put are weight, movement distance, and degree of persistence. A variable is either divided into classes or related to one or more cases. For put: weight is divided into two classes: weight of the object up to 5 kg and those over 5 kg; movement distance is divided into three classes, 10, 45, and 80, according to the distance in cm the hand is to be moved; degree of precision has two cases: to place an object *directly* or *with precision*. Each activity, consisting of its classes and cases, is assigned its standard time value based on a selected and documented motion content for the activity.

Some activities have only small norm time variations. They are treated as having no variables. Examples are *step* and *apply force*.

Each activity has a unique symbol that its standard time value is related to.

A.1.2 Supplementary Activities

Besides the two basic activities, SAM has three supplementary activities that in certain circumstances must be added to the basic activities but have no variables.

- Apply force – to apply force momentarily on an object when there is resistance in placing the object into the final position
- Step – to move the body with steps when the distance to the object or objects in a get activity or the distance to the final position in a put activity requires more than one step to support the movement
- Bend – to bend and raise the trunk of the body when the position of the object or objects in a get activity or the final position in a put activity cannot be reached from an upright body position. Note: To sit down and arise from a chair is also a bend and raise.

A.1.3 SAM Symbols for the Supplementary Activities

The symbol for a supplementary activity consists of initial letter(s) in the English word for the activity.

- Apply force (AF)
- Step (S)
- Bend (B)

A.1.4 Repetitive Activities

All manual work can be analyzed by using the two basic activities and the three supplementary activities. However, when an individual activity repeats itself identically, the deviation for the activity also repeats itself identically. The individual norm times will therefore not balance each other out and the total norm time could get a deviation that is too large.

The SAM system has, therefore, seven repetitive activities, each specially analyzed. The standard time values for these activities have small deviations and hence, can be identically repeated a number of times without risking too large of a deviation for the total norm time.

The repetitive activities are:

- screw – to rotate an object around its axis with hand, fingers, or a hand tool;
- crank – to move an object in a circular path with hand or fingers;
- to and from – to move an object in a to-and-from path with hand or fingers;
- hammer – to strike an object with a hand tool;
- read – to recognize a certain quality on a given part of an object with the eyes;
- write – to write a letter, a figure or a sign with a writing implement;
- press button – to press a button with hand or fingers.

Special repetitive activities can be developed by the individual users and added to the SAM system.

A.1.5 The SAM System Analysis Form

Fill in the top of the form with basic data for each job in order to facilitate back tracing and follow-up. Basic and supplementary activities with their variables, classes, cases, and standard time values are preprinted horizontally on the SAM sequential analysis form.

When analyzing a work method, describe complete sequences of get, put, use, and return on one line for one object or tool.

Write from left to right, never go backwards. Mark the appropriate numbers with digits.

Number each line or group of lines that belongs to one object or tool.

The frequencies (f) respectively (n) in the use column describe frequencies in increasing hierarchy.

Example: f = number of grips per screw
           n = number of screws

Use the total frequency (f) for a complete line in the "summing up" column at the far right on the analysis form.

Summarize total factors per line, multiply with the frequency (f) and note the total time.

The repetitive activity symbols and their standard time values are printed on a separate data card and should be written in the column provided on the sequential analysis form.

A.1.6 Theoretical Balance Time for the SAM System

The *activity deviation* is the deviation between the *standard time value* for a SAM activity and the *exact norm time* for the *individual* motion content for that activity.

Example: the SAM activity GS45 includes to move the hand a distance of between 10 cm and 45 cm, to grasp one object and later on to release the grasp of the object. GS45 can be carried out either with one hand or with both hands.

As the motion content of an individual GS45 deviates from the selected motion content on which the GS45 standard time value of 4 factors is based, there will be a deviation between the exact norm time for the individual GS45 and the standard time value for GS45.

To get a pencil that lies alone 10 cm away, takes of course a shorter time then to get a small screw from a box of screws 45 cm away. However, both activities are within the defined content limits for GS45 with its standard time value of 4 factors.

If, however, long and short movement distances are randomly mixed with easy and difficult grasps within the defined content limits for GS45, the deviations for the individual GS45 activities will then balance each other out. This balancing effect is achieved by the summation of all the activities in a task. For instance, an individual GS80 with a short norm time and an individual PD45 with a long norm time will balance each other out.

The SAM system has a *theoretical balance time* of 8,600 TMU, about 5 min, which is the total norm time required for the summation of SAM activity standard time values to attain a precision that would be within ±5% of the theoretically exact norm time with 95% confidence, *i.e.*, for the activity deviations to balance one another out within 5%, 19 times out of 20.

A.1.7 SAM System Activities

Basic Activities

GET – G

To gain control over one or more objects with hand or fingers content
GET begins when the hand or fingers start their movement towards the object or objects and ends when the hand or fingers have gained such a control over the object or objects that the following SAM activity can begin.

One GET can be carried out either with one hand or with both hands.

GET includes all grasp motions that are needed in order to gain control over the object or objects. GET also includes motions that release the control over the object or objects.

Variables
The time for get has two variables:

- Movement distance
- Number of objects

Movement Distance in SAM is the total distance the hand or fingers are moved in a SAM activity. If the hand is kept still and only the fingers are moved, the movement distance is then the distance that the fingertips are moved.

The movement distances are divided into three distance classes:

- Distance class 10 is movement distances from 0 cm up to 10 cm.
- Distance class 45 is movement distances over 10 cm up to 45 cm.
- Distance class 80 is movement distances over 45 cm up to the distance that can be reached with one supporting step.

These distance classes for the movement distances should always be used and be estimated.

Movement distance in get is accordingly the total distance the hand or fingers are to be moved, from the starting point of the activity to the object or objects that the hand or fingers intend to gain control over.

Number of Objects
This variable in get is related to the number of objects that are to be grasped in one get.

There are two cases: GS, to grasp a single object and GH, to grasp a handful of objects (unspecified number of objects).

On the sequence analysis form Case GS includes the time values for the distance classes for get. Hence, a GS activity includes both the movement of the hand or hands and the grasp of a single object. Case GH is designed as an addition to the GS activity on the sequence analysis form when a handful of objects are to be grasped.

Simultaneous Get
To carry out one get with one hand and simultaneously carry out another get with the other hand is two get activities, one with the distance class for the activity with the longest movement distance and the other with distance class 10. When analyzing simultaneous get activities, the type of grasp for case GS must be taken into consideration.

Case GS has got two types of grasp:

- GS with a *simple* grasp. Control over the object is gained by just closing the fingers around the object or by simply putting the hand or finger against the object.
- GS with a *complicated* grasp. Several finger motions are necessary in order gain control over the object/objects or bring the object in to the palm when several objects are grasped after each other.

For example, to take a screw from a box with one hand and simultaneously a washer from another box with the other hand, when one distance class is 80 and the other distance class is 45, is: GS80 + GS10.

If at least one of the two simultaneous get activities is a case GS with a simple grasp, GS10 should then be excluded, shown by circling it on the sequential analysis form. For example, to take a screw from a box with one hand and simultaneously a screwdriver from the table with the other hand, when one distance class is 80 and the other distance class is 45:

$$GS80 + \boxed{GS10}$$

$$PUT - P$$

To move one or more objects to a final position with hand or fingers

Final Position
The final position is the position in which the objects are planned to be placed and is the primary function of the PUT activity. The primary function for a PUT activity must therefore first be decided and then the final position can be established.

Content

PUT begins when the hand or fingers start the movement of the object or objects towards the final position and ends when the object or objects have been placed in the final position. One PUT can be carried out either with one hand or with both hands.

PUT includes, from the start of the activity to the point where the object or objects have been placed in the final position: all adjustments of the grasp, changes of the direction of the movement, stoppages in the movement and transference of the object or the objects from one hand to the other are included.

Variables

The time for PUT has three variables:

- weight;
- movement distance; and
- degree of precision.

Weight in PUT is the influence the weight of the object or objects have on the time for PUT, partly for the muscular effort in order to start the movement towards the final position and partly for the influence of weight on the speed of the movement.

Weight is divided into two classes: up to 5 kg and over 5 kg.

One AW should be added to each PUT activity when the total weight of the object(s) or the resistance to the movement is over 5 kg.

Movement distance in PUT is the total distance the hand or fingers are to be moved from the starting point of the activity to the final position. The SAM distance classes, 10, 45, and 80, should be used.

What degree of precision is required to place the object or objects in the final position?

PUT has two cases:

- PD – to place an object or objects directly.
- PP – to place an object or objects with precision.

The PD activity includes both the movement of the object or objects and the positioning of the object or objects directly at the final position. On the sequence analysis form Case PD includes the time values for the distance classes for PUT. Case PP is the precision addition to the PD activity when the object is to be placed with precision.

Type of final position must be defined *before* the decision to assign a case PD or PP activity is made.

PUT has two types of final position:

- *with insertion* of the objects into the final position, which means that the object must be aligned with the center line of the hole before it can be inserted and will result in mechanical contact between the objects;
- *without insertion* of the objects into the final position, which means to place the object in one direction *e.g.*, on a table, towards a line, corner or point.

Put with Insertion
Case PP should be assigned when force is required at the insertion or when at least one of the following five conditions appears when the object is inserted into the final position:

- Adjustment of the grasp.
- The distance between the hand and the entry position is long.
- The object is unstable or fragile.
- The entry position is concealed.
- The object must be turned right.

An insertion movement distance up to 10 cm, from the entry position to the object is fully inserted into its final position and included in the time values for PUT.

When the insertion movement distance is over 10 cm, another PUT activity with the distance class for the total insertion movement distance, including the first 10 cm, should be added to the preceding PUT activity.

Put without Insertion
Case PP should be assigned when the object must be placed in the final position without insertion and within a distance of 2 mm or when at least one of the following three conditions appears:

- The distance between the hand and the positioning point is long.
- The object is unstable.
- The final position is concealed.

Positioning Points
If a rigid object has more than one positioning point and the distances between the positioning points are not more than 10 cm, only one single PUT should be given. If, on the other hand, the distances between the positioning points are over 10 cm, each positioning point is a final position. One PUT with the distance class 10 should then be added for each additional positioning point. This rule includes both types of position.

Simultaneous Put
To carry out one PUT with one hand and simultaneously carry out another PUT with the other hand is two PUT activities, one with the distance class for the activity with longest movement distance and the other with distance class 10. For example, to place a washer with precision with one hand and simultaneously place another washer with precision with the other hand, when one distance class is 80 and the other distance class is 45:

If least one of two simultaneous PUT activities is a case PD without insertion, PD10 should then be excluded, shown by circling it on the sequential analysis form. For example, to place a washer with precision with one hand and simultaneously place a screwdriver on the table with the other hand, when one distance class is 80 and the other distance class is 45:

$$PP80 + \overparen{PD10}$$

Simultaneous Get and Put

To carry out one GET with one hand and simultaneously carry out one PUT with the other hand is one GET and one PUT with the respective distance classes for the two activities. In these situations no possible simultaneous effects are to be considered. For example, to take a screw from a box with one hand and simultaneously place a washer with precision with the other hand, when the distance class for GET is 80 and the distance class for PUT is 45:

$$GS80 + PP45$$

Supplementary Activities

APPLY FORCE – AF

To apply force momentarily on an object

It is sometimes necessary to apply force on the object in order to overcome a resistance. Apply force should then be added to the analysis. Apply force can in some situations be carried out directly after a GET.

Content

Apply force begins with a short stop in the movement, a build-up of force, sometimes together with a readjustment of the grasp; then follows the application of the force momentarily on the object. As a result of this force, a *movement of the object* might occur. This movement is either a *controlled* movement or an *uncontrolled recoil* movement.

Therefore, apply force includes a movement distance up to 10 cm. The movement is either before or after the application of force. When the movement distance is over 10 cm, a PUT with the distance class for the total movement distance should then be added to the apply force activity. Apply force can also be carried out with the foot.

Apply force shall not be used in connection with *lifting of heavy objects* (is covered by AW) or as addition to STEP.

The time for apply force has no variable.

STEP – S

To move the body, the leg or the foot.

Content

Step involves the following three types of movements:

- movement of the *entire body*;
- movement of the *leg* without moving the body; and
- movement of the *foot* without moving either the body or the leg.

One step is given each time the foot is to put down the floor or on an object. The time for step has no variable.

Step as body movement

When a movement is so long that distance class is 80, which includes one step, this not long enough; the movement distance should be supplemented with the total number of steps, including the last step before the GET activity or the PUT activity is carried out. For example, four steps have to be taken in order to grasp an object that is placed on a table. The object should then be placed directly on another table and five steps have to be taken to reach that table. Analysis:

$$4 \times S + GS10 + 5 \times S + PD45$$

The movement distance for the hand, from the moment the foot has reached the floor in the last step until the object has been grasped or placed, in the above example distance class 10, depends on where the object is located or the final position is and should therefore be estimated in each separate case.

Step as leg movement

To place the foot on a pedal, for example, and consecutively activate the pedal by moving the leg pivoted in the hip and/or the knee is one step. To then move the foot away from the pedal and place it on the floor is another step.

Step as foot movement

To put down the sole of the foot by ankle movement and then lift the sole of the foot to operate a pedal, for example, is altogether one step. If it is necessary to apply force on the pedal, an apply force activity should then be added to the step activity.

BEND – B

To bend the trunk so far that the hands reach below knee level and rise

Content

Bend begins and ends with the trunk in upright position. Bend includes a bending of the trunk forward so the hands reach below knee level and then rise to an upright position. Sometimes this is done in combination with bending of the knees and even placing one knee on the floor.

- To sit down on a chair and rise from the chair is one bend activity.
- To kneel on both knees and then rise are two bend activities.
- The time for bend has no variable.

Lifting heavy objects
When a PUT activity is being carried out and the weight of the object is over 5 kg and the movement distance is so long that body movements must be added, the object must first be lifted up towards the body with a separate PUT activity before the body movements can be carried out.

To lift up the object towards the body is the equation, AW + PD45.

For example, four steps have to be taken in order to grasp an object that is placed on a table. The weight of the object is over 5 kg. The object should then be placed directly on another table and five steps have to be taken to reach that table.

$$\text{Analysis: } 4 \times S + GS45 + AW + PD45 + 5 \times S + AW + PD45$$

It should be observed that the number of necessary steps is larger when a heavy object is moved a certain distance than when a lighter object is moved the same distance. Hence, the number of steps is determined from case to case. To walk up or down stairs or climb a ladder is analyzed as a step action. Note that the number of steps is influenced by constrains such as weight carried and other obstacles.

A.1.8 Repetitive Activities

SCREW – S

To rotate an object around its axis with hand or fingers or with a tool

Content
One SCREW activity includes a complete sequence, to rotate the object around its axis and to bring back the hand or fingers or the tool so that the following SCREW activity can start.

- To loosen or tighten a screw or a nut is a separate apply force.
- To place a screw or a nut and seat the first thread is altogether one PP activity.

When a tool is used for a SCREW activity, the tool is placed on the screw or nut with a PP activity before the first SCREW activity starts.

Variables
The time for SCREW has two variables:

- screw pattern
- thread diameter

SCREW has nine patterns:

- SA, to screw with the fingers when the resistance is so light that only finger motion is needed.
- SB, to screw with the fingers when the resistance is so apparent that both fingers' motions and hand motions are needed.
- SC, to screw with an ordinary screwdriver when the resistance is so light that only finger motions are needed.
- Note: the screwdriver may be of different types *e.g.*, blade, star, sleeve, *etc.*
- SD, to screw with an ordinary screwdriver when the resistance is so apparent that both finger motions and hand motions are needed.
- SE, to screw with a yankee driver with down and up movements.
- SF, to screw with a ratchet wrench with to-and-from movements.
- SG, to screw with a wrench by placing the wrench on the screw or screw nut in each.
- SH, to screw with an allen key by placing the key on the screw or screw nut in each SCREW activity.
- SI, to screw with a T-wrench by placing the wrench on the screw or screw nut in each SCREW activity.

In some situations a tool is used by rotating it instead of being replaced or re-gripped; a CRANK shall be assigned, not a SCREW activity.

Thread diameter

The thread diameter is valid for normal standard screws and nuts with millimeter threads and divided into four diameter classes.

- Class 1: Thread diameter up to 4 mm, Symbol 4
- Class 2: Thread diameter >4 and up to 7 mm, Symbol 7
- Class 3: Thread diameter >7 and up to 15 mm, Symbol 15
- Class 4: Thread diameter >15 and up to 26 mm, Symbol 26

Other thread types should then be compared to the closest millimeter thread and the corresponding diameter class should then be used. When carrying out SCREW activities on objects other than a standard screw or nut, *e.g.*, a screw cap on a bottle, the diameter class is half the diameter of the object at the point where the SCREW actives are carried out.

The symbol for the pattern in SCREW is written before the diameter class, for example SA15.

CRANK – CA

To move an object in a circular path with hand or fingers

Content

One CRANK includes the movement of the object as *one revolution*. When the last CRANK in a sequence of repeated CRANK activities is not a full revolution, the total number of revolutions in the sequence should be rounded off to the nearest whole number.

Example: $4.4 \rightarrow 4$ and $4.5 \rightarrow 5$

To move an object in a circular path less than half a revolution is not a CRANK but a PUT.

- movement 0.4 rev. = P (PUT)
- movement 0.8 rev. = 1 CA
- movement 1.5 rev. = 2 CA

CRANK can also be carried out with an empty hand. To move the empty hand into position for the CRANK activity is then a GET activity.

Variables
The time for CRANK has two variables:

- resistance
- precision

Resistance in CRANK is the influence the resistance has on the time for CRANK, partly for the muscular effort in order to start the movement, partly for the influence on the speed of the movement. One AW should be added to each CRANK activity when the resistance is over 5 kg. For example:

$$3 \times AW + 3 \times CA$$

Precision in CRANK is the degree of precision required at the end of the crank motion. One PP10 activity should be added to a CRANK activity when the revolution must finish within a distance of 2 mm. Also, weight allowance AW can occur.

TO AND FROM – FA

To move an object on a to-and-from path with hand or fingers

Content
One TO AND FROM includes the movement of the object in one direction and the return of the object in the opposite direction. TO AND FROM is an activity with very low control, next to instinctive. If force or care/precision is required, then activities should be analyzed as PUT. To move the empty hand into position for the TO AND FROM activity is then a GET activity. TO AND FROM can also be carried out with an empty hand.

Variables
The time for TO AND FROM has one variable: movement distance, which is the distance the hand or fingers are moved between the end points of the movements. The movement distances are divided into the three SAM distance classes.

The distance class is written after the symbol for TO AND FROM, for example, FA 45.

HAMMER – H

To strike an object with a hand tool

Content
One HAMMER includes both to lift the tool and to strike. HAMMER can also be carried out with an empty hand. To move the empty hand into position for the HAMMER activity is then a GET activity.

Variables
The time for HAMMER has one variable: case.

There are two cases:

- HA, hammer light, primarily with wrist movements
- HB, hammer heavy, primarily with forearm movements

Powerful hammering made by means of the upper arm is not considered as HAMMER but PUT.

READ – R

To recognize a certain quality on a given part of an object with the eyes

Content
READ includes only eye actions, to move the eyeballs in the direction of the object, to focus the eyesight on a given part of the object and to recognize a certain quality on that given part.

Variables
The time for read has one variable: case.

READ possesses four cases:

- RA, to read a term. One term is one word irrespective of its length or a group with a maximum of three figures and/or signs.
- RB, to compare terms and includes to read one term in one place and then read the same term in another place in order to check that both terms are identical.
- RC, to read a scale and includes to read one scale. Thus to read both the millimeter scale and the Nomi's scale on a venire are two RC. RC means analogue scales. Digital displays are red by RA
- RD, to control and includes to recognize an easy recognizable quality on an object. RD can be applied when counting objects or when determining that one has the right numbers. Note that counting is normally done in groups, *i.e.*, two by two.

NOTE – N

To write a letter, a figure, or a sign with writing implementation

Content
One NOTE includes the writing of one letter, figure, or sign with a writing implement.

Variables
The time for NOTE has one variable: case.
    There are two cases:

- NA, to print with block letters,
- NB, to write with ordinary writing.

    To place the pen into position for starting a NOTE activity is a PUT activity.
    PRESS BUTTON – PA

To press a button with a hand or finger

Content
    One PRESS BUTTON means to move the hand or finger between the buttons, to place the hand or finger on the button and to press down the button. The time for PRESS BUTTON has no variable. To move the hand into position for the first PRESS BUTTON activity is a GET activity.

APPLY FORCE in PRESS BUTTON
One APPLY FORCE should be added to the PA activity when force must be applied on the button in order to press it down. See Figures A.1 and A.2.

SAM

Time values in Factors
1 Hour = 20000 Factors
1 Factor = 5 TMU

Basic time elements

Activity	Symbol	Distance classes [cm]		
		0 - 10	(10) - 45	> 45
		10	45	80
Get Single	GS	2	4	5
Get Handful	GH	8	10	11
Put Directly	PD	2	4	5
Put with Precision	PP	5	7	8

Additional times

	Symbol	Time
Put with Weight - weight addition	AW	2

Supplementary Activities

	Symbol	Time
Apply Force	AF	3
Step	S	3
Bend down and arise	B	12
Bend Down	BD	6
Arise from Bend	AB	6

Repetitive Activities

	Symbol	Time
Hammer - per strike		
Gentle with wrist movements	HA	2
Powerful strike with forearm movements	HB	4
Read		
Read a term - per term	RA	2
Read - compare terms - per term	RB	7
Read - read a scale - per scale (analogue)	RC	8
Read - control an easily recognisable quality	RD	3
Note - per letter, figure or sign		
Note - print with block letters	NA	5
Note - write with ordinary writing	NB	3
Crank - per revolution	CA	3
Press Button - per press	PA	2

Screw

per grip with:	Symbol	Thread dimension [mm]			
		<= 4	(4) - 7	(7) - 15	(15) - 26
		4	7	15	26
Fingers - light resistance	SA	4	2	3	3
Fingers - resistance	SB	2	2	4	5
Screwdriver - ord. thread	SC	3	3	4	--
Screwdriver - self threading	SD	2	3	5	--
Yankee Driver	SE	3	3	--	--
Ratchet Wrench	SF	3	4	5	7
Wrench	SG	6	8	10	12
Allen key	SH	3	4	6	8
T-wrench	SI	6	7	8	10

To and From

	Symbol	Stroke [cm] - one direction		
		0 - 10	(10) - 45	> 45
		10	45	80
	FA	2	5	7

Bo Eklund, BE Industriutveckling 2003-03-20

Figure A.1 SAM data card (with permission from the Nordic MTM Association)

No.	Method description	GET						PUT								USE					RETURN (PUT)								SUM factors	
Object:		Date:								Dok.Id:																				
Operation:		Sign:								Drawing no:																				
		Step	Bend down	GS			Add. for Handful	Weight > 5 kilos	Step	Bend down	PD			Add. for Precision	Apply force	No. of strokes, grip et	No. of places	Time of stroke, grip et		Apply force	Weight > 5 kilos	Step	PD			Add. for Precision	Apply force	Araise		
		S	BD	80	45	10	-H	AW	S	BD	80	45	10	-P	AF	f	n	t	Kod	AF	AW	S	80	45	10	-P	AF	AB	f	Sum
																													1	
																													1	
																													1	
																													1	
																													1	
																													1	
																													1	
																													1	
																													1	
																													1	
																													1	
																													1	
																													1	
																													1	
																													1	
																													1	

Figure A.2 SAM analysis format (with permission from the Nordic MTM Association)

A.2 MTM-1 Data Cards

Permission for publishing MTM-1 and -2 data cards from the International MTM Directorate. See Figures A.3–A.5.

Reach - R

Motion Length in cm	R-A	R-B	R-C R-D	R-E	mR-A R-Am	mR-B R-Bm	m(B)	Case Description
2 or less	2.0	2.0	2.0	2.0	1.6	1.6	0.4	**A** Reach to object in fixed location, or to object in other hand or on which other hand rests.
4	3.4	3.4	5.1	3.2	3.0	2.4	1.0	
6	4.5	4.5	6.5	4.4	3.9	3.1	1.4	
8	5.5	5.5	7.5	5.5	4.6	3.7	1.8	
10	6.1	6.3	8.4	6.8	4.9	4.3	2.0	
12	6.4	7.4	9.1	7.3	5.2	4.8	2.6	**B** Reach to single object in location which may vary slightly from cycle to cycle.
14	6.8	8.2	9.7	7.8	5.5	5.4	2.8	
16	7.1	8.8	10.3	8.2	5.8	5.9	2.9	
18	7.5	9.4	10.8	8.7	6.1	6.5	2.9	
20	7.8	10.0	11.4	9.2	6.5	7.1	2.9	
22	8.1	10.5	11.9	9.7	6.8	7.7	2.8	**C** Reach to object jumbled with other objects in a group so that search and select occur.
24	8.5	11.1	12.5	10.2	7.1	8.2	2.9	
26	8.8	11.7	13.0	10.7	7.4	8.8	2.9	
28	9.2	12.2	13.6	11.2	7.7	9.4	2.8	
30	9.5	12.8	14.1	11.7	8.0	9.9	2.9	
35	10.4	14.2	15.5	12.9	8.8	11.4	2.8	**D** Reach to very small object or where accurate grasp is required.
40	11.3	15.6	16.8	14.1	9.6	12.8	2.8	
45	12.1	17.0	18.2	15.3	10.4	14.2	2.8	
50	13.0	18.4	19.6	16.5	11.2	15.7	2.7	
55	13.9	19.8	20.9	17.8	12.0	17.1	2.7	
60	14.7	21.2	22.3	19.0	12.8	18.5	2.7	**E** Reach to indefinite location to get hand in position for body balance or next motion or out of way.
65	15.6	22.6	23.6	20.2	13.5	19.9	2.7	
70	16.5	24.1	25.0	21.4	14.3	21.4	2.7	
75	17.3	25.5	26.4	22.6	15.1	22.8	2.7	
80	18.2	26.9	27.7	23.9	15.9	24.2	2.7	

Grasp - G

Code	TMU	Case Description	
G1A	2.0	**Pick-up Grasp:** any size object by itself, easily grasped.	
G1B	3.5	**Pick-up Grasp:** object very small or lying close against a flat surface	
G1C1	7.3	∅ > 12 up to ≤25 mm	**Pick-up Grasp:** interference with Grasp on bottom and one side of nearly cylindrical object.
G1C2	8.7	∅ ≥ 6 up to ≤ 12 mm	
G1C3	10.8	∅ < 6 mm	
G2	5.6	**Regrasp:** change grasp without relinquishing control.	
G3	5.6	**Transfer Grasp:** control transferred from one hand to the other.	
G4A	7.3	> 25×25×25 mm	**Select Grasp:** object jumbled with other objects so that search and select occur.
G4B	9.1	≥ 6×6×3 up to ≤ 25×25×25 mm	
G4C	12.9	< 6×6×3 mm	
G5	0.0	**Contact Grasp** (sliding or hook grasp).	

Release - RL

Code	TMU	Case Description	Code	TMU	Case Description
RL1	2.0	Normal release performed by opening fingers as independent motion	RL2	0.0	Contact release

Figure A.3 MTM-1 data card (the first of three) (with permission from the International MTM Directorate)

Move – M

Motion Length in cm	TMU					with Force/Weight			Case Description
	M-A	M-B	M-C	mM-B M-Bm	m(B)	in daN/kg up to	Static Const. SC in TMU	Dynamic Factor	
2 or less	2.0	2.0	2.0	1.7	0.3	1	0.0	1.00	
4	3.1	4.0	4.5	2.8	1.2				
6	4.1	5.0	5.8	3.1	1.9	2	1.6	1.04	**A** Move object to
8	5.1	5.9	6.9	3.7	2.2				other hand or
10	6.0	6.8	7.9	4.3	2.5	4	2.8	1.07	against stop.
12	6.9	7.7	8.8	4.9	2.8				
14	7.7	8.5	9.8	5.4	3.1	6	4.3	1.12	
16	8.3	9.2	10.5	6.0	3.2				
18	9.0	9.8	11.1	6.5	3.3	8	5.8	1.17	
20	9.6	10.5	11.7	7.1	3.4				
22	10.2	11.2	12.4	7.6	3.6	10	7.3	1.22	**B** Move object to ap-proximate or indefi-nite location, Total Clearance > 25 mm
24	10.8	11.8	13.0	8.2	3.6				
26	11.5	12.3	13.7	8.7	3.6	12	8.8	1.27	
28	12.1	12.8	14.4	9.3	3.5				
30	12.7	13.3	15.1	9.8	3.5	14	10.4	1.32	
35	14.3	14.5	16.8	11.2	3.3				
40	15.8	15.6	18.5	12.6	3.0	16	11.9	1.36	
45	17.4	16.8	20.1	14.0	2.8				
50	19.0	18.0	21.8	15.4	2.6	18	13.4	1.41	**C** Move object to exact location, Total Clearance > 12 up to ≤ 25 mm
55	20.5	19.2	23.5	16.8	2.4				
60	22.1	20.4	25.2	18.2	2.2	20	14.9	1.46	
65	23.6	21.6	26.9	19.5	2.1				
70	25.2	22.8	28.6	20.9	1.9				
75	26.7	24.0	30.3	22.3	1.7	22	16.4	1.51	
80	28.3	25.2	32.0	23.7	1.5				

Position - P

Code	Fit	Class of Fit		Symmetry	Handling	
		with secondary engage	without secondary engage		E	D
P1	Loose	No pressure required	> ± 1.5 up to ≤ ± 6.0 mm	S	5.6	11.2
				SS	9.1	14.7
				NS	10.4	16.0
P2	Close	Light pressure required	≤ ± 1.5 mm	S	16.2	21.8
				SS	19.7	25.3
				NS	21.0	26.6
P3	Tight	Heavy pressure required	Not applicable	S	43.0	48.6
				SS	46.5	52.1
				NS	47.8	53.4

Apply Pressure – AP

Code	TMU	Case Description	Components	Code	TMU	Description
				AF	3.4	Apply Force
APA	10.6	Without Regrasp	AF+DM+RLF	DM	4.2	Dwell Minimum
APB	16.2	With Regrasp	G2+APA	RLF	3.0	Release Force

Disengage – D

Code	Fit	Case Description	E	D
D1	Loose	Very slight effort, blends with subsequent move up to approx. 2.5 cm	4.0	5.7
D2	Close	Normal effort, slight recoil up to approx. 12 cm	7.5	11.8
D3	Tight	Considerable effort, hand recoils markedly up to approx. 30 cm	22.9	34.7

Figure A.4 MTM-1 data card (the second of three) (with permission from the International MTM Directorate)

Turn – T

Code	Force/Weight (daN/kg)		Time in TMU for Angular Degrees Turned										
			30°	45°	60°	75°	90°	105°	120°	135°	150°	165°	180°
T-S	Small:	≤1	2.8	3.5	4.1	4.8	5.4	6.1	6.8	7.4	8.1	8.7	9.4
T-M	Medium: >1 up to ≤5		4.4	5.5	6.5	7.5	8.5	9.6	11.6	12.7	13.7	14.8
T-L	Large: >5 up to ≤16		8.4	10.5	12.3	14.4	16.2	18.3	20.4	22.2	24.3	26.1	28.2

Body, Leg and Foot Motions

Code	TMU	Motion Length	Description
FM	8.5	up to 10 cm	**Foot Motion** pivoted at ankle
FMP	19.1		with heavy pressure
LM-	7.1	up to 15 cm	**Leg Motion** hinged at knee or hip in any direction
	0.5	each additional cm	
			Side Step lateral motion of the body
SS-C1	17.0	less than 30 cm	Use Reach or Move.
	0.2	30 cm each additional cm	Case I: complete when leading leg contacts floor.
SS-C2	34.1	30 cm	Case II: lagging leg motion must contact floor before next
	0.4	each additional cm	motion can be made.
TBC 1	18.6		**Turn Body** 45 to 90 degrees
			Case I: complete when leading leg contacts floor.
TBC 2	37.2		Case II: lagging leg must contact floor before next
			motion can be made.
B, S, KOK	29.0		**Bend, Stoop or Kneel on One Knee**
AB, AS, AKOK	31.9		**Arise from Bend, Stoop, Kneel on One Knee**
KBK	69.4		**Kneel on Both Knees**
AKBK	76.7		**Arise from Kneel on Both Knees**
SIT	34.7		**Sit**
STD	43.4		**Stand** from sitting position
W – P	15.0	per pace	**Walk**
W – PO	17.0	per pace	**Walk obstructed and/or with load > 23 kg**

Copyrighted! – Reprint not permitted! – © Copyright 1955 ... © 2008
MTM Association for Standards and Research

International MTM Directorate
info@mtm-international.org

MTM-1
Data Card
(SI – metric system)

Do not attempt to use this chart or apply Methods-Time Measurement in any way unless you understand the proper application of the data. This statement is included as a word of caution to prevent difficulties resulting from misapplication of the data.

	Time Units			
	TMU	seconds	minute	hour
The time values in this data card are equivalent to a performance of 100 % LMS	1	0.036	0.0006	0.00001
	27.8	1	1	-
	1,666.7	-	1	-
	100,000	-	-	1

Simultaneous Motions

Motions not included in above table:
T Turn: normally easy with all motions except when Turn is controlled or with Disengage must be analyzed
AP Apply Pressure : each case must be analyzed
P3 Position : always difficult
D3 Disengage : normally difficult
RL Release : always easy
D Disengage: any class may be difficult if care must be exercised to avoid injury or damage to object

= Easy to perform simultaneously.
= Can be performed simultaneously with practice
= Difficult to perform simultaneously even after long practice. Allow both times.

W: within the area of normal vision
O: outside the area of normal vision
E: easy to handle
D: difficult to handle

Eye Travel and Eye Focus

Code	TMU	Description
ET	15.2 × T/D	**Eye Travel** T: distance between points from and to which the eye travels D: perpendicular distance from the eye to the line of travel T
	maximum 20.0 TMU	
EF	7.3	**Eye Focus**

Figure A.5 MTM-1 data card (the fourth of four) (with permission from the International MTM Directorate)

A.3 MTM-2 Data Card

See Figures A.6 and A.7.

Figure A.6 MTM-2 data card (the first of two) (with permission from the International MTM Directorate)

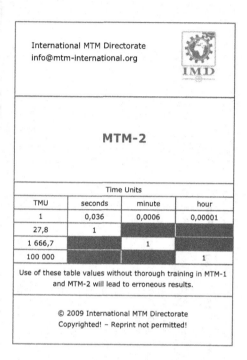

Motion Length in **cm**	0 - ≤5	>5 -≤15	>15-≤30	>30-≤45	>45
Distance Range	**5**	**15**	**30**	**45**	**80**

Get						
Case of Get	**Code**	Distance Range **5** (TMU)	**15**	**30**	**45**	**80**
No grasping motion	**GA**	3	6	9	13	17
One grasping motion	**GB**	7	10	14	18	23
More than one grasping motion	**GC**	14	19	23	27	32
Weight / Force	**GW**	1 TMU per 1 kg/daN for weights/forces ≥ 2 kg/daN per hand				

Put						
Case of Place	**Code**	Distance Range **5** (TMU)	**15**	**30**	**45**	**80**
No correction	**PA**	3	6	11	15	20
One correction	**PB**	10	15	19	24	30
More than one correction	**PC**	21	26	30	36	41
Weight / Force	**PW**	1 TMU per 5 kg/daN for weights/forces ≥ 2 kg/daN per hand				

Complementary Motion Sequences	**Code**	**TMU**
Apply Pressure	**A**	14
Regrasp	**R**	6
Crank	**C**	15
Eye Action	**E**	7
Foot Motion	**F**	9
Step	**S**	18
Bend and Arise	**B**	61

Simultaneous Motions

Case	**GA**	**GB**	**GC**	**PA**	**PB**	**PC**
PC	with practice	difficult	difficult	with practice	difficult	difficult
PB	with practice	with practice	difficult	with practice	with practice	
PA	easy	easy	with practice	easy		
GC	easy	easy	difficult			
GB	easy	easy				
GA	easy					

2 **PB** can be performed simultaneously, with practice, in the area of normal vision, as long as the "POSITIONS" are symmetrical.

= easy	= with practice	= difficult
easy to perform simultaneously	can be performed simultaneously with practice	difficult to perform simultaneously even after long practice; allow both times

Figure A.7 MTM-2 data card (the second of two) (with permission from the International MTM Directorate)

A.4 Motion Economy

A.4.1 Principles of Motion Economy as Related to Use of the Human Body

1. The two hands should begin as well as complete their motions at the same time.
2. The two hands should not be idle at the same time except during rest periods.
3. Motions of the arms should be made in opposite and symmetrical directions and should be made simultaneously.
4. Hand and body motions should be confined to the lowest classification with which it is possible to perform the work satisfactorily.
5. Momentum should be employed to assist the worker wherever possible, and it should be reduced to a minimum if it must be overcome by muscular effort.
6. Smooth continuous curved motions of the hands are preferable to straight-line motions involving sudden and sharp changes in direction.
7. Ballistic movements are faster, easier, and more accurate than restricted (flexible) or " controlled" movements.
8. Work should be arranged to permit an easy and natural rhythm wherever possible.
9. Eye fixation should be as few and as close together as possible.

A.4.2 Principles of Motion Economy as Related to Use of the Work Place

1. There should be a definite and fixed place for all tools and materials.
2. Tools, materials, and controls should be located close to the point of use.
3. Gravity feed and containers should be used to deliver material close to the point of use.
4. Drop deliveries should be used wherever possible.
5. Materials and tools should be located to permit the best sequence of motions.
6. Provisions should be made for adequate conditions for seeing. Good illumination is the first requirement for satisfactory visual perception.
7. The height of the work place and the chair should preferably be arranged so that alternate sitting and standing at work are easily possible.
8. A chair of the type and height to permit good posture should be provided for every worker.

A.4.3   Principles of Motion Economy as Related to the Design of Tools and Equipment

1. The hands should be relieved of all work that can be done more advantageously by a jig, a fixture, or a foot-operated device.
2. Two or more tools should be combined whenever possible.
3. Tools and materials should be prepositioned whenever possible.
4. Where each finger performs some specific movement, such as in typewriting, the load should be distributed in accordance with the inherent capacities of the fingers.
5. Levers, hand wheels, and other controls should be located in such positions that the operator can manipulate them with the least change in body position and with the greatest speed and ease (Barns 1949).

A.5   Work Sampling

The definition of work sampling is as follows: "A work sampling study consists of a large number of observations taken at random intervals. In taking the observations, the state or condition of the object of study is noted, and this state is classified into predetermined categories of activity pertinent to the particular work situation. From the proportions of observations in each category inferences are drawn concerning the total work activity under study."

Let's first introduce sample size effect in work sampling. Figure A.8 shows a sampling result with a different sample number and size. In this manner, holes set as a sample unit and back chart of the "i" letter is a complete normal distribution. A difference of sample numbers makes the difference to see through the shape of a normal distribution. For instance, the increased sample number of 30 means better identification of a normal distribution as a letter of "i" than 15.

Standard deviation is a quick reference point for testing any observed distribution for normality. The formula for determining the sample size for a confidence level of 68%, or 1 sigma, is:

$$S_p = \sqrt{\frac{p(1-p)}{N}}$$

where S = desired relative accuracy
S_p: Standard deviation, desired relative accuracy
p: percentage expressed as a decimal
N: number of random observations (sample size)

In the normal curve, the area enclosed between $\pm 1\sigma$ is 68.26%, $\pm 2\sigma$ is 95.45%, and $\pm 3\sigma$ is 99.73%.

The formula for a confidence level of 95% and accuracy of ±5% is as follows:

$$S_p = 2\sqrt{\frac{p(1-p)}{N}}$$

The definition of occurrence curve consists of average and standard deviation.
Distribution of sample averages will become more and more compact as the sample size increases.

A.5.1 Calculation of Sampling Sizes

Work sampling is a tool that helps realize present practice based on the laws of probability theory. Sampling method can save study time and cover wide areas in a study. It is an efficient method to know a certain subject practice in an economical amount of time. A feasibility study for productivity can be used as a convenient study. See Figure A.8.

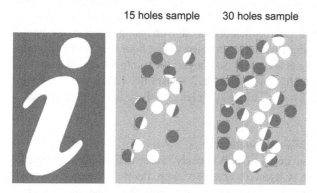

Figure A.8 Sampling size and facts image

There are a few practical components of facilitating a work sampling study.

- Keep the necessary number of observations based on theoretical calculation.
- Keep randomness when setting observation times.
- Ensure a clear definition of classified observation items.

Keep the Necessary Number of Observations Based on Theoretical Calculations.
The number of total observations is calculated as follows. The formula for a confidence level of 95% and accuracy of ±5% is as follows:

$$S^2 p^2 = 4\left[\frac{p(1-p)}{N}\right] = \frac{4p(1-p)}{N}$$

Further, to calculate N where $p = 25\% = 0.25$, and $S = \pm5\% = \pm0.05$:

$$N = \frac{4p(1-p)}{0.0025p^2} = \frac{4(1-p)}{0.0025p} = \frac{1600(1-p)}{p}$$

$$N = \frac{1600(1-0.25)}{0.25} = 4800$$

In the practice of work sampling study, S, the desired relative accuracy is recommended as 5%. The remaining 95% gives a confidence result on a sampling based on the background of normal distribution. Two sigma, or two standard deviations, is 95.45%; about 95% of data confidence, but not in the remaining 5%. One sigma is 68.27; three sigmas equals 99.73.

The observation term is recommended as at least one week. Observation results reflect the difference of days in a week. Observation sample size means the number of observation timing multiplied by the number of observation objects that are observed during observation time.

Keep Randomness When Setting Observation Times.
There are two methods for observation: *fixed interval* and *random timing of observation*. The observation number is the same, but to keep representing the whole facts, fixed interval observation cannot guarantee facts at a certain level of confidence.

Random sampling times can be demonstrated with using a telephone book. Open the pages and three-digit numbers are used as the hour (the first digit) and minute (last two digits). Digits are 0 to 9, so convert them into 8 h and 60 min. For example, p.329 is 2:18(3 × 8 h = 24: 2 o'clock, 2 × 60 min = 12: 10 min, 9 × 9 min = 81:8 min, so 2 o'clock 18 min). This is sufficient, as there is no need for precision in this case.

Clearly Define Classified Observation Items.
When planning observation items of a WS study, clear definitions and simple expressions are imperative. Observers come to shop floors to study facts and round up different tasks for follow-through within a short time. Therefore, observers must decide on observation items very quickly. Note this list of corresponding classification and observation items for an FM work study:

- instruction – methods, set-up, preparation that explains performance target and operation order;
- supervising – measuring, writing memos, watching, shop floor meetings, measuring operators' work time, evaluate memos regarding workers;
- communication – speaking, telephone calls, writing;
- desk work – operating computer in-house and externally;
- movement – materials handling;
- meeting – review performance of others;
- extra work – direct operation, help set up operators, repair machines; and
- absence – cannot find in FM's own shop (see Figure A.9).

	classification	observation items	contents
1	instruction	instruction	methods, set-up, preparation explaning performance target operation order
		ordering	shop meeting
2	supervising	measuring	measuring operators' work time
		writing memo	memo regarding workers
		watching	look around FM's shop
3	communication	speak	
		telephone	
4	desk work	writing	
		operating PC	
5	movement	in own shop	
		out of own shop	
		materials handling	
6	meeting	reg. performance c.	
		others	
7	extra work	direct operation	direct operation instead of workers
		help set up operators reparing machines	these are not normal work of FM
8	absence		can not find in FM's own shop

Figure A.9   Work sampling observation items: FM activities

A.6 25% Selection

Allowed time values for MDC models are selected as 25% selection methods. Average or mean values are suitable measurements of the time value of WU, but 25% selection methods are recommended because of new design methods that are currently taught by foremen and industrial engineers. These methods are points to be instructed on because time values are dependent on skills such as labor performance matters. Time values are a subordinate issue for implementing new methods. This is why labor performance control is recommended. Figure A.10 illustrates this method: two distributions show whether adequate instruction of methods has been given or not.

Also, 25% selected value is the mean of distribution based on nonadequate instruction of present methods. The left-hand observation results are the time study results based on nonadequate instruction of reset conditions. The time values required for MDC WU are time value based on adequate instruction of new methods, and prospects can be acquired with 25% selection of current time study results.

The procedure to find allowed time for MDC design methods follows.

Total observation number (15) × 25% = 3.75 = 4. This 4 means a time value that is the fourth of accumulated occurrence distribution from the least time value 0.28 min. That is 0.32 min. This 0.32 min is selected as the allowed time value of MDC WU. See Figure A.11.

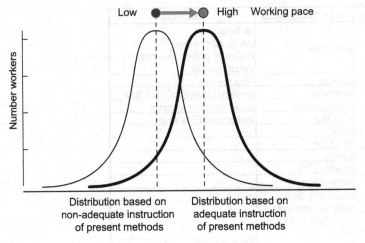

Figure A.10   To prospect mean value of based on instruction

Figure A.11   Allowed time value
through 25% selection

observed results	
no.	time (min)
1	0.40
2	0.35
3	0.30
4	0.28
5	0.35
6	0.33
7	0.35
8	0.34
9	0.32
10	0.30
11	0.38
12	0.33
13	0.35
14	0.34
15	0.39

distribution	
occurrence	time (min)
1	0.28
2	0.30
1	0.32
2	0.33
2	0.34
3	0.35
1	0.36
1	0.38
1	0.39
1	0.40

References

Barnes R (1949) 16 principles of motion economy as first stated by the Gilbreths, 1923 as "A fourth dimension for measuring skill for obtaining the one best way." Soc Indust Engin Bull 5:174–236

Barnes R (1980) Motion and time study, design and measurement of work, 7th edn. Wiley, New York

Heiland R, Richardson W (1957) Work sampling. McGraw Hill, New York

Mundel M (1978) Motion and time study improving productivity, 5th edn. Prentice-Hall, Upper Saddle River, NJ

Bibliography

Antis WH, Honeycutt JR, Koch EN (1973) The basic motion of MTM, 4th edn. The Maynard Foundation & Prentice-Hall, Upper Saddle River, NJ

Burnham DC (1972) Productivity improvement. Columbia University Press, New York

Carrol P (1954) Time study for cost control. McGraw-Hill, New York

Fried HO, Knox Lovell CA, Schmidt S (2008) The measurement of productive efficiency and productivity growth. Oxford University Press, New York

Fujita A (1953) Basics of industrial engineering. Kenpakuya, Tokyo, Japan

Gadiesh O, Gilbert JL (1998) Profit polls: A fresh look at strategy. Harvard Business Review, May–June 1998

Herbert S (1971) The meaning and measurement of productivity. Bureau of Labor Statistics Bulletin 1714

Honeycutt A, William JM, Kock EN (1973) The basic motion of MTM, 4th edn. The Maynard Foundation & Prentice-Hall, Upper Saddle River, NJ

IMD, International Institute for Management Development (1997), (1998), (1999), The World Competitiveness Yearbook, Lausanne, Switzerland

Institute of Industrial Engineers (1983), Industrial Engineering Terminology, Institute of Industrial Engineers, Norcross, GA

Juran JM (1995) Managerial breakthrough: The classic book on improving management performance. McGraw-Hill, New York

Kadota T, Sakamoto S (1992) Chapter 55: Methods analysis and design. In: Salvendy G (ed) Handbook of industrial engineering. Wiley, New York, pp. 1415–1445

Krick EV (1965) An introduction to engineering & engineering design. John Wiley & Sons, New York

Lokiec M (1977) Productivity and incentives. Bobbin Publications, Los Angeles, CA

Mali P (1978) Improving total productivity. Wiley, NY

Meadow DH, Meadow DL, Randers J, Behrens WW III (1972) The limit to growth. Universe Books, New York

Morony MJ (1964) Facts from figures. Penguin Books, New York

Morrow RL (1957) Motion economy and work measurement. The Ronald Press, New York

Mundel M, Danner D (1994) Motion and time study improving productivity, 7th edn. Prentice-Hall, Upper Saddle River, NJ

Nalebuff B, Brandenburger AM (1996) Co-opetition. Harper Collins Business, London, UK

Polanyi M http://infed.org/thinkers/polanyihtm

Prentice-Hall, Upper Saddle River, NJ

Prokopenko J (1987) Productivity management, a practical handbook. International Labour Office, Geneva, Switzerland

Riggs JL, Felix GH (1983) Productivity by objectives: Results-oriented solutions to the produc-
  tivity puzzle. Prentice-Hall, Englewood Cliffs, NJ
Sakamoto S (1977) How a Japanese firm doubled productivity without capital investment. Inter-
  national Productivity Conference, Sydney, Australia
Sakamoto S (1977b) Japanese firm doubles productivity. Institute of Practitioners In: Work
  study, organization and methods. Management Services, UK
Sakamoto S (1981) Practices of industrial engineering. Kenpakusya, Tokyo, Japan
Sakamoto S (1983a) MOP: A head of OA, adopt IE to office. Annual Industrial Engineering
  Conference, Louisville, KY
Sakamoto S (1983b) Practices of work measurement. Japan Management Association, Tokyo,
  Japan
Sakamoto S (1985a) MDC engineering manual. Japan Management Association, Tokyo, Japan
Sakamoto S (1985b) MOP: Managing Office Productivity. Japan Management Association,
  Tokyo, Japan
Sakamoto S (1989) Process design concept. Ind Eng 3:31–34
Sakamoto S (1990) Really high Japanese productivity. Japan Management Association, Tokyo,
  Japan
Sakamoto S (1991a) The MDC training manual. Productivity Partner Inc, Nara, Japan
Sakamoto S (1991b) MDC: Productivity engineering methods. Japan Management Association,
  Tokyo, Japan
Sakamoto S (1992a) Design concept for methods engineering. In: Hodson WK (ed) Maynard
  industrial engineering handbook. McGraw Hill, New York
Sakamoto S (1992b) A practical manual of MDC. Japan Management Association, Tokyo, Japan
Sakamoto S (1997) Japanese firm doubles productivity, Management Services, Institute of Prac-
  tioners in Work Study, Organization and Methods
Sakamoto S (2002) A study of company dignity. Toyokeizai Shinposya, Tokyo, Japan
Sakamoto S (2006) Methods design concept: An effective approach to profitability. J Philippine
  Ind Eng
Sakamoto S (2007) Productivity management: Innovative approach for white color. Sangyou
  Nouritsu University, Tokyo, Japan
Sakamoto S (2009) Return to work measurement. J Indust Engng 3:24
Schonberger RJ (1986) World-class manufacturing. The Free Press, New York
Skinner W (1978) Manufacturing in the corporate strategy. Wiley-Interscience, Hoboken, NJ
Slywotzke AJ, Morrison DJ (1997) The profit zone. Times Business, New York
Stockholm Environment Institute (1996) Sustainable economic welfare in Sweden: A pilot index
  1950–1992. Stockholm Environment Institute, Stockholm, Sweden
Swedish Federation of Productivity Services (1993) SAM training program. Swedish Federation
  of Productivity Services, Stockholm, Sweden
Taylor FW (1911) The principles of scientific management. Harper, New York
Tiefenthal R (1975) Production: An international appraisal of contemporary manufacturing
  systems and the changing role of workers. McGraw-Hill, New York
von Weizsäcker EU, Lovins AB, Lovins LH (1995) Faktor Vier. Rocky Mountain Institute,
  Boulder, CO
Zandin KB (1980) MOST work measurement system. Marcel Dekker, New York

Index